The Biotechnology Business

The Biotechnology Business

A strategic analysis

Peter Daly

Frances Pinter (Publishers) Ltd
Rowman & Allanheld Publishers

© Peter Daly 1985

First published in Great Britain in 1985 by
Frances Pinter (Publishers) Limited
25 Floral Street. London WC2E 9DS

Published in the United States of America in 1985 by
Rowman & Allanheld Publishers.
(A Division of Littlefield. Adams and Company)
81 Adams Drive. Totowa. New Jersey 07512

British Library Cataloguing in Publication Data
Daly. Peter
    The biotechnology business: a strategic analysis
    1. Biotechnology industries—History
    I. Title
    338.4'76606   HD9999.B442

    ISBN 0-86187-551-6

Library of Congress Cataloging in Publication Data
Daly. Peter. 1953–
    The biotechnology business

    Bibliography: p.
    Includes index.
    1. Biotechnology industries. 2. Market surveys.
I. Title [DNLM: 1. Biomedical Engineering.
2. Industry. 3. Technology. Medical. QT 34 D153b]
HD9999.B442D35 1985 338.4'76208 85–10877

    ISBN 0-8476-7460-6

Typeset by Folio Photosetting. Bristol
Printed in Great Britain by Biddles Ltd. Guildford

Contents

List of tables

Acknowledgements

The author wishes to acknowledge the following sources for permission to reproduce copyright material:

Biotechnology News for Table 3.2
Dr. Andrew Pickett, Porton International Ltd., for Table 4.1
Genex Corporation for Table 4.2
Financial Times for Table 9.3

I am indebted to two major sources. Firstly, Michael Porter of Harvard University, whose work on industry structural analysis and competitive strategy has been widely influential. Secondly, to the biotechnology report of the Office of Technology Assessment (US Congress) which is the most authoritative source of data available on the biotechnology industry. Finally, I would like to thank my family and friends for their support and encouragement during the preparation of the manuscript.

1 Introduction

1.1 Background

The use of biological processes dates back into antiquity when brewing, cheese making and baking were discovered. The microbiological bases of these processes were only elucidated following developments in the science of biology in the eighteenth and nineteenth centuries.

In the early twentieth century, biological processes were used to produce commodity chemicals such as acetone and butanol but these fermentations were largely superseded with the large development of the petrochemical industry. After the Second World War, a number of biotechnology-based industries became established producing substances including antibiotics, amino acids, enzymes, etc. In the period from 1953 (when the structure of DNA was discovered) to the present there has been an explosion in our knowledge of the molecular basis of biological systems. This enormous increase in new knowledge has also led to the development of powerful new techniques with industrial applications. Two of the most important of these are recombinant DNA technology (genetic engineering) and hybridoma technology, which were discovered between 1970 and 1975.

The commercial significance of these discoveries was soon appreciated in the United States and in the period 1976 to the present many new companies were formed to exploit the new technology. Established companies in the chemical, pharmaceutical, food and oil industries followed the new start-ups in investing in biotechnology R & D. New and established companies in all the major industrial countries also began to invest in biotechnology.

Over the period 1976 to the present, biotechnology has attracted considerable public interest especially in the United States. The first major public interest was in the area of safety. This followed the Asilomar Conference in 1975, when leading American molecular biologists called for public debate on the possible dangers of genetic engineering research. These 'dangers' were subsequently found to be largely insubstantial and the strict guidelines which had been adopted in the late 1970s were relaxed in most countries. Currently, public controversy in the United States is centred on the issue of deliberate release of genetically engineered organisms into the environment for such purposes as crop protection.

The involvement of financial institutions in supplying venture capital and the subsequent public share offerings of the new companies also aroused public and media interest. Governments have responded to this situation by commissioning reports on the state of national biotechnological competitiveness and by investing in various initiatives. Most OECD countries currently have policies to promote the development of biotechnology.

Biotechnology, unlike some other areas of high technology, is not a single technology but a group of technologies including existing bioprocessing technology and new technologies such as recombinant DNA and hybridoma technology. Its applications are multisectoral, as in the case of microelectronics. Many scholarly texts and books for the layman exist on the various technologies and their application areas. It is not the intention here to replicate this material but background data is supplied to clarify the analysis of industry trends and strategies.

This work is concerned with the 'new biotechnology' industry, its emergence and the strategic contraints and options facing the firms that are involved in it. Chapters 2 and 3 provide some background data on the applications of biotechnology and the companies attempting to exploit it world-wide.

The emergence of the biotechnology industry may be seen as being shaped by four major groups of factors. Firstly, there are the market characteristics of those areas within which biotechnological innovation is occurring. Secondly, and related to this, are those characteristics that are found in any emerging industry such as embryonic companies, early entry barriers, and technological uncertainty. The special relationship between 'science' and 'technology' is also of major relevance both for the structure of the industry world-wide and the strategic behaviour of the firms involved. Finally, the role of government is assessed, as national governments see biotechnology as one of the 'sunrise industries' and it receives considerable attention from industrial planners wishing to enhance national competitiveness.

Having examined the major forces shaping the industry, I examine the performance of a number of major companies to date. These case studies analyse the financing and strategy of some of the leading new biotechnology firms (NBF's)* and consider the involvement of multinational enterprises and the relation of their 'biotechnology strategies' to their overall strategic goals. Following the case studies, a typology of biotechnology strategies is presented and the benefits and risks of each is assessed. The final chapter is an analysis of some of the major factors which have an impact on success in the emerging industry. Previous work concerned with success and failure in industry is related to biotechnology and leads to the identification of the key success factors.

Before considering the application areas of biotechnology and the companies involved, it may be useful to review briefly the main technologies which constitute biotechnology.

1.2 The technologies

Genetic engineering (recombinant DNA)

Genetic engineering or recombinant DNA technology is the technique of introducing hybrid DNA containing genes of interest into organisms in order to make the organisms produce a product of interest such as an enzyme, hormone or other protein. The organisms used as host are generally micro-organisms such as *Escherichia coli,* *Bacillus subtilis* or yeast, although some proteins are being produced from genetically engineered mammalian cells. The application of recombinant DNA to the production of a particular substance involves a number of steps. The new gene is introduced into the host through use of a suitable vector, such as a plasmid or virus. It is necessary to construct a high level expression system to produce maximum amounts of the desired product. The laboratory process, based on the organism produced by recombinant DNA technology, must be scaled up to pilot plant and manufacturing plant and the appropriate recovery and purification operations designed[See References 22,23,100]. While genes for cloning can be obtained by excision from the original genome, it is often convenient to synthesize the gene chemically once its structure is known and then use this for cloning[3].

---

* The term 'new biotechnology firm' (NBF) was first used in the report of OTA, *Commercial Biotechnology:- An International Analysis* [99].

Genetic engineering or recombinant DNA technology has facilitated the industrial mass production of proteins. Substances previously too scarce are now available in sufficient quantities for clinical trials. Genetic engineering is also the basis of process innovations reducing the cost of existing fermentation products such as enzymes and amino acids. Recombinant DNA technology can also be applied to the genetic improvement of plants and in due course to animals.

Bioprocessing

Bioprocessing involves the conversion of a raw material substrate into a product using microbial fermentation or enzymes. It is also concerned with the recovery of a product, its separation from the fermentation broth and purification[98,99].

Existing bioprocesses are used in the production of antibiotics, enzymes, amino acids and other specialty chemicals. With the development of recombinant DNA technology, much new research activity has gone into developing improved bioprocessing systems for the production of recombinant DNA products on an industrial scale. Bioreactors may be batch or continuous. In batch operations a biocatalyst is added to the substrate in the fermenter and the process is allowed to proceed to completion before emptying the reactor and starting purification. Most industrial bioprocesses are batch processes. In continuous processing, substrate and nutrients are continually fed into the bioreactor, processed and the outlet stream is continuously treated to purify the product. The biocatalyst, either microbial cells or enzymes, may be immobilized on beads or other solid supports: this technique allows the biocatalyst to be re-used (thus saving costs) and avoids contamination of the product by the biocatalyst.

Much R & D effort is being put into the automation of bioprocesses including the development of continuous sensor devices and the interfacing of process control with on-line computers.

Hybridoma technology

Antibodies are proteins produced in vertebrates in response to foreign proteins or substances. Since a number of antibodies are produced in response to any given protein, conventional antisera consist of mixtures of antibodies. Hydridoma technology allows the production of highly specific antibodies from single clones of cells[25,98]. These are termed monoclonal antibodies (Mabs). The process of producing Mabs involves immunization of a laboratory animal with the antigen. Following this, the animal is killed and the spleen which contains antibody-producing B lymphocytes is obtained. These cells are then fused with cancerous myeloma cells to produce hybrid cells or hybridomas. Using appropriate selection procedures, a hybridoma clone producing the antibodies to the particular antigen is obtained and these cells can then be used to obtain larger quantities of the Mabs either by growing them in laboratory animals, or by cultivating them *in vitro*. The hybridoma cells have some of the properties of both parent cells; from the B lymphocytes they obtain the ability to produce antibodies and from the myeloma they obtain the ability to replicate indefinitely.

Monoclonal antibodies with their high degree of specificity have a number of important uses including:

- incorporation into *in vitro* diagnostic systems such as new tests against hepatitis B or cancers;
- use in *in vivo* diagnostic imaging for detection of metastasized tumour cells;
- therapeutic uses including immunization and immunotoxins, i.e., the linking of cytotoxic drugs to specific antibodies to create targetable drugs against tumour cells;
- tissue typing for use in surgery and blood typing;
- use in purification and separation of biological molecules.

A few companies have developed technology for large-scale production of monoclonals *in vitro*. The leading company in this area is probably the new British company Celltech which produces Mabs in a 1,000-litre air uplift reactor. Another leading contender is Damon Biotech in the United States which has developed a proprietary encapsulation process for growing hybridomas.

Protein engineering

Protein engineering is a technology at an early state of development but it will be very important in the development of the biotechnology industry and will be used in conjunction with genetic engineering. It involves the modification of protein structure to improve the function of proteins or to design entirely new proteins[116]. Enzymes involved in industrial processes, such as starch degradation or the manufacture of corn syrup, could be modified so as to improve their tolerance of temperature or alter pH optimum or other characteristics. Such a modification could have important consequences on the economics of the process. Another possible application is in therapeutic proteins, which could be redesigned to improve their biological activity. Another possibility, suggested by Kevin Ulmer [112] of Genex Corporation, is the alteration of the recognition specificity of DNA-binding proteins including repressors and restriction endonucleases, important biochemical tools in recombinant DNA technology.

The development of protein engineering will depend on a number of other techniques including X-ray diffraction methods, computer molecular modelling and chemical synthesis of DNA. The latter area is already well developed and will allow the easy construction of genes coding for the modified proteins. X-ray diffraction allied to computer modelling and interactive graphics will elucidate the rules governing protein structure and its relationship to function.

In this book the term 'second generation products' is used to refer to analogues of natural products produced by synthetic oligonucleotide synthesis and subsequent recombinant DNA methods. Such oligonucleotides may be designed using protein engineering. 'First generation' products are natural substances produced by conventional recombinant DNA technology.

Bioinformatics

As biotechnology develops in the areas both of research and of industrial application, information technology and microelectronics will become increasingly important. This area of convergence between information technology and biotechnology is termed bioinformatics [86]. It covers diverse fields including the use of computers in protein engineering as described above, and the use of special software for DNA sequence analysis including artificial intelligence programmes for simulating cloning experiments. The American company Intelligenetics has developed software for use by molecular biologists in cloning experiments. This software allows such abilities as the 'translation' of DNA sequences into corresponding amino acid sequences, the identification of restriction sites, search for homology and the simulation of cloning. The DNA sequence data bases in Europe and the United States will be available for interaction by 'intelligent terminals' used for running

Intelligenetics-type software. The importance of this technology in speeding up research was demonstrated in 1983 when a major discovery in the area of oncogenes (genes implicated in cancer) was made by comparing sequence data with that held on previously discovered genes in the data bases. An increasing trend in the future will be the automation of the research process itself. Automated DNA synthesizers capable of rapidly manufacturing genes or oligonucleotides are already in use in industry and acedemia. Further developments here will include the interfacing of many items of laboratory equipment with computers and the increasing use of artificial intelligence for the planning of research strategies. These capabilities will mean a very large increase in the speed at which research can be carried out and in the rate of increase of new knowledge.

Another important area of bioinformatics is automated process control in the fermentation-based industries. Such developments imply that biotechnology will not be a labour intensive industry.

2 Applications of biotechnology

The applications of biotechnology are very broad and affect a number of sectors including pharmaceuticals, specialty chemicals, food, agriculture and commodity chemicals. This chapter presents an outline of the main application areas, as a background for readers unfamiliar with the subject.

2.1 Pharmaceuticals and diagnostics

Biotechnology will have a profound impact on the pharmaceutical industry and an in house capability in biotechnology will be necessary for pharmaceutical companies in the future. The applications of genetic engineering and hybridoma techniques combined with a greatly increased understanding of the molecular causes of disease will lead to a transformation of medicine. Many diseases for which adequate therapy does not exist at present will become more amenable to successful therapy and diagnosis, including cancers, inflammatory diseases, viral diseases, neural and mental disorders and parasitic infections. The effect of biotechnology will be to transform medicine from an empirically based discipline, to one based on an understanding and design of molecular processes. These greatly enhanced capabilities in health care have major social considerations for populations both in the industrialized world and the underdeveloped world. However, such considerations are beyond the scope of this work.

The applications of biotechnology to the pharmaceutical sector are considered under a number of headings including:

- drugs using recombinant DNA technology
- lymphokines (including interferons)
- drug targeting
- vaccines
- applications to conventional drug production
- new diagnostic technologies

8

Drugs using recombinant DNA technology

Recombinant DNA technology (genetic engineering) allows the potential production of any protein found in the body, previously obtainable only in minute quantities. In practice only proteins of a particular size would be manufactured using recombinant DNA; it would be more economical to synthesize short proteins (peptides) chemically.

A wide range of proteins of potential therapeutic value are currently being developed (Table 2.1). These include proteins involved in the immune system such as interferons for cancer

Table 2.1 Some proteins with pharmaceutical applications being developed through use of genetic engineering

Protein	Application
Alpha interferon	Treatment of cancers, colds
Gamma interferon	Treatment of cancers
Interleukin-2	Cancer therapy
Human growth hormone (HGH)	Growth promotion, burn treatment
Calcitonin	Therapy of bone disease
Alpha 1-antitrypsin	Treatment of emphysema
Endorphins	Analgesia
Other neuropeptides	Memory therapy, depression, Parkinson's disease and senility
Human serum albumin	Treatment of shock and burns
Factor VIII	Anti-haemophilic factor
Tissue plasminogen activator	Breakdown of blood clots

Source: OTA Report [99]

therapy, proteins currently obtained from blood (including human serum albumin and Factor VIII), hormones and neuroactive peptides. As well as proteins intended for therapeutic use, proteins which will form the basis of improved vaccines are also being developed through use of recombinant DNA technology. These include human vaccines against, for example, hepatitis B, viral diseases, parasitic diseases (in the future), and animal diseases such as scours.

The effect of this technology is to reduce the cost of substances previously available only at considerable cost and in minute quantities. The new technology allows these substances to be tested for clinical efficacy and subsequently to be manufactured. For instance, a genetically engineered organism grown in a 450-litre fermenter was able to supply in a few batches enough human growth hormone to fulfil British demand which had previously been met by extracts from the pituitary glands of 60,000 cadavers.

Lymphokines, interferons and anti-cancer substances

The genetically engineered drugs which have attracted most attention to date are lymphokines, which includes interferons. Lymphokines are substances produced by lymphocytes or white blood cells and play an important role in the immune response. Consequently, such substances are being studied for their potential use in the treatment of cancers and viral infections. Interferons are currently undergoing clinical trials in various countries but are not yet approved (February 1985) by the Food and Drug Administration (FDA) for full-scale marketing in the United States. Interferons are produced in genetically engineered *E. coli* and also in yeast. Much early excitement was generated by these substances but their clinical usefulness is still not clear and a variety of other potentially commercially competing anti-cancer lymphokines and other substances are also under development. Chief among these is interleukin-2 (IL-2) which promotes the growth of T-cells and may be useful also in treating the acquired immune deficiency syndrome (AIDS). Other lymphokines may also have potential industrial importance including B cell growth factors (BCGF) and B cell differentiation factors (BCDF). Other approaches to cancer therapy rely on substances which attack or modify cancer cells rather than modulate the immune response. Lymphotoxin for instance, recently cloned by Genentech, is claimed to attack cancer cells selectively while a tumour necrosis factor is also being investigated for possible therapeutic use in tumour regression.

Drug targeting

The use of monoclonal antibodies to direct cytotoxic drugs against specific antigens on cancer cells is also under investigation. A number of years of further development are required before products are developed. This technology would allow cytotoxic drugs such as ricin or cobra venom to be selectively delivered to the sites of cancer metastases within the body. The technology could also be used for diagnostic imaging *in vivo* and a number of companies are pursuing this line of R & D.

Vaccines

The new technology will also allow the development of improved and safer vaccines (especially for viral disease) to replace existing vaccines as well as entirely new vaccines. Instead of using a purified preparation of surface antigen from blood infected with the virus. producing vaccines will in future involve bulk production of the relevant antigens either through rDNA-based fermentation or mammalian cell culture. At present. a variety of new vaccines are under development including those for hepatitis B. and various animal diseases including foot and mouth disease. Improved bacterial vaccines will also be introduced for diseases such as cholera and tuberculosis.

Applications to conventional drug production

Biotechnology may also be applied to the production of conventional drugs such as antibiotics to improve processing. The genes coding for the rate-limiting enzyme step in a biosynthetic pathway may be cloned to increase the enzymatic activity and hence antibiotic production within the relevant micro-organism. Genetic engineering may also be used to enhance manufacturing in other ways such as producing solvent-extractable antibiotics. Since all existing microbial processes used in the industry are theoretically open to improvement using molecular techniques. it can be expected that a high proportion of such processes will be improved over the next two decades.

New diagnostic technologies

Many new diagnostics are appearing: changes include new developments within the existing immunodiagnostics markets and also the development of entirely new diagnostic technologies. Within the existing immunodiagnostics market. the trend is towards greater use of enzyme immunoassay (EIA) and away from radio immunoassay. This is because EIA does not require handling of hazardous radioactive substances. Over the past few years dozens of diagnostic kits incorporating Mabs (rather than the conventional polyclonal) have been introduced to the market both by new biotechnology firms (NBFs) and large established firms. In many instances, such as reproductive hormones, thyroid function and sexually transmitted disease, these new kits are competing with existing assays based on polyclonal antibodies. In other cases, such as for various cancers, the new Mab kits are new assays not previously available. The market for Mab *in vitro* diagnostics is currently $5 – 6m in the United States but is predicted to grow to $300–500m by 1990. This is currently the fastest growing area of the biotechnology industry and the one with by far the largest number of new product introductions

Another important development in diagnostics is the area of DNA probes. These are pieces of DNA which are homologous to the DNA being detected and will thus hybridize with it. Through use of suitable labels (radioactive or biotin) the hybridization can be detected. DNA probes may be used in the diagnosis of infectious disease and the detection of prenatal genetic defects. This technology may compete with Mab-based diagnostics in certain cases.

2.2 Specialty chemicals

Existing specialty chemicals made using fermentation technology, include a wide variety of compound types; enzymes, amino acids, microbial polysaccharides and microbial pest control agents. The applications of new biotechnology are leading to several types of innovation:

• the application of genetic engineering to the fermentation production of existing specialty chemicals thus improving the economics of production;
• the application of protein and molecular engineering to the modification of existing specialty chemicals or development of new ones with enhanced functional characteristics.

The first form of innovation, i.e. the application of genetic engineering, is already occurring with a variety of recombinant specialty chemicals under development. The latter area requires much more fundamental research before product innovations can be envisaged and is thus more long term. So in the near future the application of biotechnology to specialty chemicals will result mainly in process improvements leading to specialty chemicals that will compete with conventionally produced chemicals on the basis of price.

Enzymes

Enzymes are protein molecules which catalyse chemical reactions in cells. They are used industrially in a wide variety of processes including the manufacture of detergents, sweeteners, cheese making and medical products. Genetic engineering is being applied in two main ways for the production of enzymes. Existing microbial systems for making enzymes may be modified by use of recombinant DNA, firstly, to make it more efficient and, secondly, where enzymes are extracted from plant or animal tissue, the genes coding for those enzymes may be cloned in micro-organisms thus allowing production through fermentation. An example of the latter is the microbial cloning of chymosin, which has been achieved by a number of companies. Chymosin is used in making cheese and is currently obtained from calf stomachs.

Longer term developments here include the identification and commercialization of new enzymes for industrial use and the modification of the structure of enzymes to improve their activity.

Amino acids

Amino acids are used in food and animal feedstuffs to enhance flavour and provide nutritional supplements. They also have pharmaceutical applications in enteral solutions. The present market is dominated by a number of Japanese producers, the largest of which is Ajinomoto. Biotechnology is being applied to improve the process economics of amino acids already made by microbial processes, such as glutamic acid, and also to develop microbial processes for other amino acids.

Microbial polysaccharides

These include thickening and flocculating agents and lubricants which are made by microbial methods. The main polysaccharide polymers in use at present are xanthan gum and dextran. Xanthan gum is used for gelling purposes in food and may also find application in enhanced oil recovery. Biotechnology developments here will lead to the production of new polymers. However, since the biochemistry and genetics of the producing micro-organisms is poorly understood it will be some time before recombinant DNA technology can be applied to biopolymer production.

2.3 Food

Biotechnology will affect existing bioprocesses within the food industry, lead to new enzymatic and microbial processes and in some instances to new food products and ingredients. Many industries already use enzyme and microbial technology including brewing, distilling, baking, dairy products, meat products and confectionery. These processes may be improved through the application of recombinant DNA technology. The application of immobilized enzyme technology and protein engineering will lead to new enzymatic processes for food flavour, stability, etc.

The application of biotechnology will also lead to new food ingredients and new foods. The new sweetener aspartame is finding wide use in other foods. Further development of single cell protein technology may lead to the use of single cell protein (SCP) for human food: in Britain Rank Hovis McDougall are developing a fungal SCP for human use. The food industry will also benefit from the applications of plant biotechnology such as the ability to produce tomatoes with low water content which are consequently easier to process for tomato sauces and other products.

2.4 Agriculture

Applications of biotechnology to agriculture can be considered under two areas:

- animal health care and reproduction
- plant production and protection

Animal health care and reproduction

Recombinant DNA technology and hybridoma technology are generating many new animal health care products. These include vaccines for scours, foot and mouth disease, bovine papilloma, etc. In the longer term molecular studies of parasitology should lead to production of the vaccines for parasitic diseases of domestic animals and rapid tests for diagnosis of these conditions. Apart from vaccines, a range of other animal pharmaceutical products will be available such as bovine somatostatin to improve milk production, growth hormones and bovine interferons.

Embryo manipulation technologies will lead to more efficient and productive animal husbandry and it is likely that recombinant DNA technology will be applied to improve features of animal physiology and biochemistry of economic importance such as milk production, size, meat quality etc.

Plant production and protection

Micropropagation technology allied to genetic manipulation of plants should result in rapid production of plants with desirable traits such as herbicide, pest and drought resistance and perhaps in the longer term the ability to fix atmospheric nitrogen. These developments will have a significant effect on agriculture in the longer term. Improved diagnosis of plant diseases will result from the current research efforts into areas such as DNA probes. Improved plant protection will also result from the applications of biological control including microbial insecticides with genetically engineered toxicity and the use of pheromones.

2.5 Commodity chemicals

The commodity chemical industry is currently undergoing a structural reorganization resulting from overcapacity in the United States and Europe and increased global competition from the Middle East, Mexico and South East Asia. Although a range of commodity chemicals could theoretically be produced by biotechnological methods, it is still more economic to use petroleum feedstock and the rate of substitution of biological for chemical processes depends to a large extent on petroleum feedstock availability and price. It is unlikely that biotechnological feedstocks and processes will replace many existing chemical processes and alternative feedstocks from coal liquefaction, shale, etc. may be more competitive than use of biological feedstocks. Biotechnology is not expected to have a significant impact on the commodity chemical industry until about the year 2000 or after.

3 Biotechnology companies

3.1 The United States

New biotechnology firms (NBFs)

The new biotechnology firms (NBFs) are new entrepreneurial start-ups formed generally since 1976 with the assistance of venture capital. Over one hundred of these firms were founded between 1976 and 1985. Many of the founders or co-founders were academic scientists wishing to exploit their expertise commercially in areas such as recombinant DNA technology.

Typically such firms are research intensive with a high proportion of staff involved in R & D (Table 3.1) and do not manufacture their

Table 3.1 Numbers of Ph.D. level research staff within some NBFs, 1982–1983

Company	Total no. employees	No. Ph.D.s within research staff
Amgen	100	45
California Biotechnology	44	21
Chiron	67	44
Collaborative Research	125	25
Genex	219	54 (approx.)
Integrated Genetics	125	25

Source: Company annual reports and prospectuses

own products in their early years. Initial R & D expenses are financed by venture capital, funds raised through public stock offerings and other financial mechanisms, in addition to income generated from contract R & D and product licensing. Many companies show

Table 3.2 Comparative financial data on selected public NBFs, 1983 ($'000)

Company	Revenue	Expenses	Net income	Total assets
Amgen*	4,347	7,826	(3,479)	55,438
Biogen	18,437	29,453	(11,664)	111,428
Biotechnica International	698	2,953	(3,650)	—
California Biotech†	5,266	4,986	205	17,185
Cambridge Bioscience	535	1,521	(986)	5,125
Centocor	7,407	8,243	(836)	24,604
Genentech	47,003	45,537	1,128	48,744
Genex	11,091	16,470	(5,379)	52,107
Hybritech	15,965	16,439	(474)	48,066
Integrated Genetics	3,046	5,217	(2,171)	27,755
Molecular Genetics	6,915	6,463	(451)	—

* Nine months.
† 30 November 1983.
Source: Biotechnology News. Vol. 4. No. 14. 1 June 1984. Reproduced by permission.

significant losses on operations at this stage (Table 3.2). While income is generated from contract R & D and licensing, the initial financing of the company and subsequent refinancing, is effected in a variety of ways including venture capital, public stock offerings and R & D limited partnerships. These are described more fully below.

The strategy of most NBFs (either implicit or explicit) is to transform themselves from research companies into fully integrated manufacturing and marketing companies. This process of transformation, referred to here as 'forward integration', involves the commercialization of R & D, investment in plant and the financing of entry into product markets including capital investments and cost of regulatory barriers. While no NBF has yet achieved total forward integration, a number of leading companies are at present in a period of transition in which product sales represent an increasing proportion of total operating income. During this period the NBF devotes increasing resources to its own internally generated R & D and reduces the proportion of research resources involved in projects for other clients. The research clients of the NBFs have included a range of major chemical, pharmaceutical, food and energy corporations many of which have equity investments in the NBFs as described below.

The original 'Big Four' NBFs included Genentech, Cetus, Biogen and Genex. While Genex and Genentech are fairly advanced in forward integration, Cetus and Biogen are further from becoming manufacturing entities. Other leading NBFs include Centocor, Chiron, Amgen, Enzo-Biochem, Molecular Genetics, Damon Biotech, Collaborative Research, Hybritech and Integrated Genetics (Table 3.3).

Table 3.3 Leading new biotechnology firms (NBFs) and their areas of activity

Company	Activity areas
Amgen	rDNA indigo, chicken growth hormone, consensus IFN
Biogen N.V.	Alpha, gamma-IFN, interleukin-2
Bio-Response Inc.	Large-scale cell culture
Biotech Research Laboratories Inc.	Mabs, restriction enzymes, plant biotechnology
Calgene Inc.	Plant biotechnology
Cambridge Bioscience Corp.	Viral and bacterial diagnostics animal vaccines
Centocor Inc.	Cancer diagnostic assays, Mab imaging and therapeutics
Cetus Corp.*	Vaccine for scours, lab equipment, CMV test, beta-IFN, interleukin-2
Collaborative Research Inc.	Kidney plasminogen activator, EMIA diagnostics, research products, IL-2 and rDNA rennet
Collagen Corp.	Collagen implants
Creative Biomolecules Inc.	DNA linkers, adapters, probes
Damon Biotech Inc.	Patient-specific Mabs, large-scale cell culture
Ecogen Inc.	Microbial pesticides

Table 3.1 (*continued*)

Company	Activity areas
Endorphin Inc.	Neuroactive peptides
Enzo Biochem Inc.	DNA probes, interferon
Genetic Design Inc.	DNA synthesizers
Genentech	rDNA drugs; human insulin, growth hormone, tPA, calcitonin, Factor VIII, lymphotoxin
Genetic Engineering Inc	Animal embryo sex determination
Genetic Systems Corp.	Mabs for infectious disease, tissue typing, cancer diagnostics
Genetics Institute Inc.	Lymphokines, Factor VIII, tPA, biological control, DNA probes
Genex Corp.	Aspartic acid, phenylalanine, enzyme cleaner, enzymes and vitamins and other amino acids
Hybritech Inc.	Mab diagnostics, *in vitro* cancer tests and imaging
Immunex Corp.	Cancer therapeutics; IL-2, MAF, BCGF, IL-1
Immunomedics Inc.	RIA-based cancer detection
Integrated Genetics Inc.	DNA probe diagnostics, hepatitis B vaccine, tPA
Intelligenetics	Software for molecular biology research
Molecular Genetics Inc.	Therapeutic veterinary Mabs
Ribi Immunochem Research Inc.	Pesticides, frost control products, embryo manipulation
Xenogen Inc.	DNA probes, IFN delivery systems, feed additives

* Agracetus activities are not included
Source: Genetic Engineering News, Vol. 4, No. 8, 1984 and company reports and prospectuses

Of the above companies, case studies are presented on Genentech, Genex, Cetus and Centocor in Chapter 7.

NBFs are carrying out R & D in a wide range of areas: new human drugs including cancer therapeutics, diagnostics, animal healthcare, animal and plant agriculture and specialty chemicals. In general NBFs emphasize either recombinant DNA technology or hybridoma technology. Companies such as Genentech and Genex are applying recombinant DNA technology to the production of pharmaceuticals and specialty chemicals respectively [58,64]. On the other hand companies such as Hybritech and Centocor are applying hybridoma technology to develop new diagnostic tests [29]. The latter area has shown the most rapid new product development of any area of the 'new biotechnology'. Forty-one monoclonal antibody based kits had been approved for use in the United States by summer 1983 [99].

Many NBFs are pursuing R & D into a number of high-value products such as lymphokines (Table 3.4) and there is considerable rivalry to get a product on the market first. Some product areas such as interleukin-2 are becoming crowded and it is unlikely that all of the companies listed in Table 3.4 will be able to compete successfully in the market place.

Since their formation the NBFs have pioneered a number of major technical achievements in biotechnology including:

- the use of recombinant DNA technology to produce therapeutically useful proteins;
- the use of recombinant DNA technology to produce specialty chemicals;
- the appliation of hybridoma technology to diagnostics and the development of large-scale cell culture both for hybridomas and mammalian cells;
- scientific/technical advances in new hosts and expression systems, immunology and molecular studies of cancer and cloning in plant cells.

These technical achievements have not yet been translated into commercial success. The difficulties and opportunities facing NBFs in this area are discussed in later chapters.

Financing of new biotechnology firms (NBFs)

NBFs have been financed through a number of different financial mechanisms in the United States. The chief mechanisms include:

Table 3.4   Examples of companies carrying out R & D on the same products

Product	Companies
Tissue plasminogen activator	Genetech
	Biogen
	Integrated Genetics
	Chiron
	Collaborative Research
Alpha interferon	Genentech
	Amgen
	Biogen/Schering-Plough
	Enzo Biochem
	Genex
	Burroughs-Wellcome
	Cetus
Interleukin-2	Biotech Research Labs
	Cetus
	Chiron
	Ajinomoto
	Cellular Products
	Du Pont
	Otsuka Pharmaceutical
	Fujisawa Pharmaceuticals
	Green Cross
	Suntory
	Hayashibara
	Immunex
	Biogen
	Genetics Institute
	Quidel
	Genex
	Interferon Sciences
	Israel Institute for Biological Research
	Biomedical Institute
Factor VIII	Genetics Institute
	Biogen
	Genentech
	Chiron
	Integrated Genetics
Tumor necrosis factor	Biogen
	Celltech/Sankyo
	Genentech
	Smith Kline

Source: OTA Report [99]

- venture capital;
- public stock offerings;
- R & D limited partnerships.

Venture capital

Most NBFs obtained their initial start-up financing from venture capital sources. Venture capital can be obtained from a number of sources including funds managed by venture capital firms and corporate venture capital. Venture capital funds obtained from insurance companies, corporate pension funds, private institutions and wealthy individuals have been used to finance both the start-up and subsequent phases of the early NBF. Public venture capital funds make it possible for the individual small investor to put money into NBFs. An example of this is Biotechnology Investments Ltd., which was formed by N & M Rothschild for the specific purpose of investing in biotechnology companies. Biotechnology Investments Ltd. has assets of over $60m and has invested in Integrated Genetics, Immunex, Amgen and the leading British NBF, Celltech. Another mechanism by which the small investor can invest in biotechnology companies is through high technology mutual funds.

Corporate venture capital has also been an important source of finance for NBFs in their early stages and many of the leading established firms involved in biotechnology have equity investments in NBFs (Table 3.6).

In 1980, nine NBFs received first-round venture capital financing in the United States. This rose to twenty-nine in 1981 but then dropped to thirteen in 1982. In 1980 there were also five second-round financing arrangements with nine in 1981 and seventeen in 1982*. Valuations of NBFs have decreased over the period 1980–1982. Data presented by the United States Office of Technology Assessment shows that NBF valuations in 1980 were $5–25m for 25% of the company, but after 1982 ranged from $2 to $4 for 40–50% of the company[99]. This coincided with a decrease in prices of stocks which is discussed below.

Public stock offerings

Bio-Response was the first NBF to 'go public' in October 1979 and raised $3.3m in its initial offering. This was followed in 1980 by the

---

Bio/Technology, September 1983

first public stock offering of Genentech which attracted much public attention as share prices went from the initial price of $35 to $89 within the first hour. In 1981 Cetus raised $120m in its initial offering. a situation unlikely to be matched by an NBF in the foreseeable future. In all. about two dozen NBFs have now had their first public stock offering.

By 1982 stock prices of NBFs had fallen and new NBFs found it considerably more difficult to raise the expected amount on their stock offerings. Thus Molecular Genetics obtained only $3.3m which represented less than one-third of its goal. In the first half of 1983 this situation improved. but, by late 1983, companies were delaying initial stock offerings in the hope of an upturn in the new issues market. Companies which went public at this time raised less than expected sums. The initial prospectus of Advanced Genetic Sciences filed with the Securities Exchange Commission in June 1983 included a goal of two million shares at $20–30 per share. However, the company came to market in September 1983 with a 750,000 share offering at $15 per share. Similarly, Biotechnology General Corporation offered 800,000 shares at $13 after initially planning to offer one million at $17–20.

Many stocks have fallen considerably since their offering. Amgen shares were offered at $18 in June 1983 and fell to $6 by November 1983 and $5 by May 1984. The fall in the price of shares and the difficulty experienced by later firms in obtaining public finance are due to a number of factors. Firstly, the market for equities in high technology companies showed a decline which also influenced biotechnology investments. Investors became more familiar with the technology between 1980 and 1982 and there was some degree of disillusion with biotechnology stocks as it was realized that there was a long lead time in commercializing biotechnology R & D and consequent long lead time on return on investment. Many of the earlier valuations of NBFs were excessive and the movement of stock prices has reflected a more realistic appraisal of their value. One biotechnology analyst, Glenn Knickrehm [82]. states that the movement of stock prices subsequent to the initial public offering depends on the relationship between the initial public offering price and the market to R & D ratio (market value to R & D budget) for the industry average:

if the initial public offering price is lower than that indicated by the industry average market to R & D ratio the price often moves up. Conversely, if the initial public offering price is higher than that indicated by industry averages, the price may fall.

He presents two examples of this: Molecular Genetics and Genex experienced a price increase relative to industry (specialty chemicals) over time and had a low initial public offering, while Biotechnica International and Advanced Genetics Systems suffered price declines relative to the industry and had high prices at the initial public offering. This reinforces the argument that many NBFs were initially over-valued.

Research and development limited partnerships

Research and development limited partnerships (RDLP) are a mechanism by which investors put money into specific projects rather than the company itself. The investors become limited partners and get a first-year write-off against taxable income and R & D tax credits. The limited partners can receive a percentage of royalties if the product is successful and this can also qualify for favourable tax treatment with the profits taxed at the capital gains rate (20 per cent) rather than income (50 per cent). The general partner, which is the sponsoring company or a subsidiary manages the venture. The RDLP has ownership of any technology developed and the sponsoring company must buy out the limited partners to exploit the discoveries. This is an attractive financing mechanism for NBFs because it allows them to raise large sums to finance R & D or clinical trials without surrendering equity and R & D expenses can be kept off their balance sheets. For the investors the RDLP acts as an attractive tax shelter as well as a potentially profitable source of income. RDLPs did not attract much attention before the mechanism was used by biotechnology companies, as the amounts raised were small. Biotechnology companies have raised large sums using RDLPs: Genetech raised $89m in its first two RDLPs in 1982 and 1983 and a further $30.7m in late 1984. One innovative use of the RDLP has been its application to fund clinical trials.

The RDLP as a financial mechanism is less successful now than in 1982–83 due to a number of factors. Firstly, there has been a general trend in equities and the somewhat depressed state of investment in biotechnology generally. The market for these specialized investments may also be limited and there are many RDLPs on offer. These trends have contributed to a situation where firms have raised less

Table 3.5 Major R & D limited partnerships in biotechnology.
         1981–84

Company	Amount ($m)	Year
Agrigenetics	55.0	1981
California Biotechnology	27.5	1982
Genetic Systems	3.4	1982
Hybritech	7.5	1982
Molecular Genetics	11.1	1982
Neogen	1.0	1982
Genentech	55.0	1982
Cetus	75.0	1983
Alza	16.0	1983
Genentech	34.0	1983
Serono Labs	29.0	1983
Xoma	16.0	1983
Biogen	60.0	1984
Genentech	30.7	1984

Source: Bio/Technology. Vol. 2. No. 8. August 1984

than expected. Another significant factor is a recent change in the tax laws which make RDLPs less attractive to investors.

American established companies

Major pharmaceutical, chemical, food and energy corporations are involved in biotechnology through a number of mechanisms including:

- in-house investment in R & D and plant;
- licensing and marketing arrangements with NBFs;
- investment and linkages with universities;
- acquisition of NBFs.

Investment in R & D & manufacturing plant
Establshed firms have devoted large sums to in-house R & D and manufacturing plant since 1980. Eli-Lilly has invested $40m in manufacturing plant for human insulin and $60m in a new biotechnology research centre. DuPont spent $120m on biotech-

nology in 1982, while Pfizer has recently opened a $80m life science complex. Schering-Plough is investing $100m in plants to manufacture, purify and formulate its alpha interferon product [99].

Equity participation in NBFs

Established firms in pharmaceuticals, chemicals, energy etc. have taken equity in NBFs both at the start-up phase and in subsequent refinancings. In some instances their investments represent joint development projects while in others the established company merely has a passive investment. Established firms frequently have contract research done on their behalf by those NBFs they have equity in. Some of these investments may be seen in Table 3.6.

Table 3.6   Some equity investments in new biotechnology firms by established firms in the United States

NBF	Established firm	Date
Collaborative Research	Dow	1981
Genetic Systems	Cutter Laboratories	1983
Genentech	Fluor	1981
	Lubrizol	
Immunorex/Centocor	FMC	1981
Enzo Biochem	Johnson & Johnson	1982
Agrigenetics	Kellogg	1982
Genex	Koppers	1979
Biogen	Schering-Plough	1978, 79
Collagen	Inco	1980,81
	Monsanto	1980
Genencor/Genentech	Corning Glass	1981
	A E Staley	1984

Source: OTA Report [99]

Licensing and marketing arrangements

Many of the first generation of biotechnology products result from the licensing to established firms of technology from NBFs. Schering-Plough's alpha interferon ('Intron') resulted from the licensing of the recombinant DNA technology from Biogen while Eli-Lilly's recombinant human insulin ('Humilin') resulted from technology obtained from Genentech. Established firms have also engaged in a wide variety of marketing arrangements with products developed by NBFs. Celltech's diagnostic kits are being distributed in Japan by

Sumitomo while Abbott is marketing cancer diagnostic kits developed by Centocor.

Investments in universities
Major firms have invested in university research in a number of ways and for differing reasons. Some investments have been for the purpose of 'gaining a window' on new technology and for training corporate scientists. Others such as that of Monsanto are joint research projects while other companies have contracted universities to carry out specific research on their behalf (Table 3.7).

Table 3.7 Some corporate investments in universities and research institutes in the United States

Company	Institution	Amount	Purpose
Hoechst	Massachusetts. General Hospital	$70m over 10 years	Establishment of 100 person molecular biology dept. Training of Hoechst personnel
DuPont	Harvard Medical School	$6m	Research on molecular genetics
Exxon	Cold Spring Harbor Lab.	$7.5m over 5 years	Training of Exxon scientists in molecular biology
Corning Glass Eastman Kodak Union Carbide	Cornell University	$2.5m each over 6 years	Establishment of Biotechnology Institute
W.R. Grace	Massachusetts Institute of Technology	Up to $8.5 over 5 years	Genetic engineering (applied)
Monsanto	Washington University	$23.5m over 5 years	Co-operative research projects in biotechnology

Source: Biotechnology News

Acquisition of NBFs
The major firms have acquired a number of NBFs as part of their technology acquisition strategy. DNAX, which had expertise in

Table 3.8　Some biotechnology company acquisitions in the United States

Company acquired	Technology area	Acquired by
DNAX	rDNA. drug delivery	Schering-Plough
Phytogen	Plant biotechnology	J.G. Boswell
Cetus Madison	Plant biotechnology	W.R. Grace
Agrigenetics	Plant biotechnology	Lubrizol
Atlantic Antibodies	Mab immunoglobulin purification	Charles River Laboratories
Bioclinical	Diagnostics	Ventrex Lab.
Immuno Modulators Laboratories	Immuno-modulators	Ventrex Lab.
Biotherapy Systems	Cancer diagnostics and therapeutics	Damon Biotech
ICL Scientific (diagnostics)	Diagnostics	Hybridoma Sciences

Sources: *Biotechnology News; Business Week* (11 June 1984); *European Chemical News* (10 September 1984); OTA Report [99]

recombinant DNA and new drug delivery systems was acquired by Schering-Plough and acts as a research department of that company. Major companies including Pfizer, Stauffer, Upjohn and Occidental Oil purchased seed companies during the 1970s. In recent years some established firms have acquired NBFs which were focused on plant biotechnology. Examples of this include the acquisition of Phytogen by J.G. Boswell and of Agrigenetics by Lubrizol (Table 3.8). Acquisitions of NBFs by major corporations are likely to increase significantly in the late 1980s.

3.2 Japan

The commercialization of biotechnology in Japan relies almost exclusively on established firms in the pharmaceutical, chemical and food industries. Japan does not possess the venture capital system and entrepreneurial climate which led to the establishment of the American NBFs. Biotechnological developments include R & D and new product introduction within companies with existing bioprocesses and also diversification by companies into areas of biotechnology in which they have no previous experience, e.g., food companies diversifying into biotechnology-based pharmaceuticals.

Leading Japanese companies include Ajinomoto, Kyowa Hakko Kogyo, Snow Brand, Kirin, Sapporo, Toray, Takeda, Suntory (Table 3.9). Japanese companies are strong in the technical areas of fermentation, quality control, and bioengineering scale-up and purification. These are crucial technologies required for the construction and operation of full-scale manufacturing plant in biotechnology. These technical strengths are likely to be a considerable advantage to Japanese companies in international competition. Japanese companies also have considerable experience in microbial processes involved in making antibiotics, specialty chemicals etc., and they dominate a number of product markets such as amino acids. Until the early 1980s Japanese companies lagged considerably behind American companies in expertise and R & D activity in molecular biology. However, since 1981 the situation has changed considerably. The Ministry of International Trade and Industry (MITI) and the Science & Technology Agency (STA) have co-ordinated a major entry into this field by Japanese companies and technology transfer from the United States also appears to be important. Although expenditure by Japanese companies on bio-technology R & D is significantly smaller than in the United States, recent Japanese achievements in molecular biology are impressive (Table 3.10). The United States is, however, still dominant in this area.

3.3 Europe

United Kingdom

Biotechnology companies in the United Kingdom include both established pharmaceutical and chemical companies and also NBFs. The United Kingdom has more NBFs than any other country except the United States. The leading NBF is Celltech which was founded in 1980 as a partnership between private and public venture capital. Celltech is involved in Mab-based diagnostics and molecular biology and is a world leader in large-scale hybridoma production. Celltech has linked up with Boots to market its products and has formed a joint venture Boots-Celltech Ltd. In 1984, the Agricultural Genetics Company was established with the assistance of the state-

Table 3.9 Some leading Japanese companies involved in biotechnology

Company	Activity
Ajinomoto	Production of amino acids, R & D into anti-cancer substances and immunomodulators
Kyowa Hakko	Production of amino acids, foods, pharmaceuticals, chemicals. R & D into specialty chemicals, pharmaceuticals including anti-cancer (γIFN), tPA
Suntory	Alcoholic beverage production. R & D into interferons
Kirin Brewery	Alcoholic beverages, R & D into anti-cancer agents
Toyo Jozo	Alcoholic beverages, pharmaceuticals, R & D into immunosuppressants
Snow Brand	Food products, R & D into anti-cancer agents, erythropoietin and plant biotechnology
Otsuka Pharmaceutical Co.	R & D into rDNA in the United States, textiles, pulp production
Toray Industries	R & D into interferons
Hayashibara Company	Gamma interferon R & D and production
Kikkoman	Production of soya sauce, R & D on enzymes, diagnostics, drugs and plant biotechnology
Sumitomo Chemical	Production of chemicals, pharmaceuticals, R & D into monoclonal antibodies, diagnostics and interferon
Meiji Seika Kaisha	Food processing. R & D into antibiotics, interferon

Sources: Genetic Engineering News, November/December 1984; *Bio/Technology,* April 1984; *Japan Bio Industry Letters,* April 1984

Table 3.10 Some recent Japanese achievements in molecular biology. 1983–1984

Institution/company	Activity
Takeda	Interleukin-2 introduced into clinical trials
Osaka University	Recombinant DNA on Salmonella type
Yamasa Shoyu	Method for microbial identification based on rDNA technology
Snow Brand Milk Products	Mab method for extracting erythropoietin from urine
Kyowa Hakko/Cancer Research Institute	Production by rDNA of protein binding leukaemia viruses. Potential use in vaccine
Kikkoman Shoyu	Development of vector to cause *E. coli* to make 25% of cellular protein as cloned gene product
Hayashibara Biochemical Lab. Ltd.	Initiation of phase 1 clinical trials on carcino breaking factor
Green Cross	Cloning and sequencing of gene for human urokinase
Green Cross	Production of Mabs to tetanus toxin from human-human hybridomas
Mitsubishi Chemical Industries Nagoya University	Development of method for excretion of cloned protein in *E. coli*
Fujisawa Pharmaceutical Co.	Mass production of somatomedin C in *E. coli* using rDNA
Taisho Pharmaceutical Co./ Institute of Applied Microbiology. University of Tokyo	Development of a host-vector system for cloning in mammalian cells

Sources: Abstracts in Biocommerce. cumulative. editions 1–6; *Japan Bioindustry Leters.* April 1984

backed British Technology Group to exploit research performed by the Agricultural Research Council. Other NBFs in Britain include Inveresk Research, Cambridge Life Sciences, Monotech Laboratories, IQ Bio and Imperial Biotechnology.

Established firms which have invested in biotechnology in the UK include ICI, Burroughs Wellcome, G.D. Searle, Glaxo and Beecham [21]. ICI commissioned a full-scale plant for the production of single cell protein in 1980 and is continuing R & D into this area as well as into microbial polymers. Burroughs Wellcome has set up a subsidiary to exploit genetic engineering technology and is investing in large-scale cell culture technology for production of pharmaceutical proteins [28]. Wellcome Biotechnology offers the service of scaling-up cell production systems developed by other companies in return for a license to produce and market the product. Wellcome is carrying out R & D into improved foot and mouth disease vaccines and a malaria vaccine. Glaxo has traditionally been an antibiotic company but is now expanding its R & D into biotechnology applications to arthritis and cardiovascular disease. This company is experiencing significant market penetration in the United States for its anti-ulcer drug. Recombinant DNA technology may also be used in the production of antibiotics at Glaxo. G.D. Searle has also invested in the biotechnological production of drugs and the artificial sweetener, aspartame.

Federal Republic of Germany

Venture capital is virtually non-existent in Germany and consequently there has been little entrepreneurial activity in biotechnology [99, 113]. German corporate involvement in biotechnology comes almost exclusively from major chemical and pharmaceutical companies. Germany possesses some of the leading chemical and pharmaceutical companies (Hoechst is the world's largest pharmaceutical company) and is therefore potentially in a strong position to exploit biotechnology. However, German companies have been very slow to invest in the new technology and to form links with universities. This situation is now changing following the controversy over Hoechst's $70m investment in Massachusetts General Hospital in order to train its scientists. Major companies involved in biotechnology R & D include Bayer AG, Schering AG, Boehringer Ingleheim, Behringwerke and Boehringer Mannheim. Schering AG is establishing a $33m institute for recombinant DNA research in association with the State of Berlin. BASF and the University of Heidelberg are establishing a joint programme for training and basic research. Some companies have also formed links with American NBFs; Schering, for example, has agreements with Genetics Institute (Boston) and Genex (Rockville, Maryland). The strong market position of German companies in markets affected by biotechnology is an important asset but the late entry and conservatism demonstrated by these companies is a major disadvantage in international competition.

France

Industrial biotechnology in France is represented by established firms in the pharmaceutical and food industries and by a few NBFs.

Major companies involved in biotechnology R & D include Rhône-Poulenc, Elf-Aquitaine and Roussel-Uclaf [113]. Elf-Aquitaine and two other companies. Crédit Agricole and Lafarge Coppe established the Agritech biotechnology fund in the United States in 1983 to transfer technology back to the sponsoring companies. There are a number of NBFs including Transgène, and Groupement de Génie Génétique; Transgène was established in 1980 by collaboration between the Institut Pasteur and Strasbourg University, Paribas, a consortium of investment bankers and the manufacturing firms of Institut Merieux and Roussel-Uclaf. Transgène is carrying out R & D into interferons and a rabies vaccine. Groupement de Génie Génétique is an industrial subsidiary of a number of research institutes including the Institut Pasteur.

Netherlands

The Netherlands has a number of major companies investing in biotechnology including Akzo-Pharma N.V., Intervet International and Gist Brocades[113]. The latter company is one of the leading enzyme manufactures in the world. Intervet International (a subsidiary of Akzo) was the first company to market a vaccine made through recombinant DNA technology. In late 1984/early 1985 two leading United States NBFs, Centocor and Molecular Genetics, announced their intention to establish subsidiaries in the Netherlands.

Switzerland

Switzerland possesses three major multinational pharmaceutical companies, Hoffmann-La Roche, Sandoz and Ciba-Geigy as well as a number of smaller firms [99, 113]. Ciba-Geigy has made a major commitment to entering the biotechnology industry and has invested $19.7m in a biotechnology research centre in Basle, Switzerland. The Company is also investing in a $7m biotechnology research centre in the United States (Research Triangle Park, North Carolina). Hoffmann-La Roche has significant investments in biotechnology-based cancer drugs. It has developed alpha interferon (licensed from Genentech) and is researching other lymphokines. Sandoz is investing in research into lymphokines, neurobiology and other areas and has research contracts with a number of American research centres as well as a $5m investment in the NBF Genetics Institute (Boston). The Swiss American company, Biogen, ws one of the pioneers of the application of recombinant DNA technology to pharmaceuticals and holds a patent on the recombinant DNA production of alpha interferon.

Sweden

Companies investing in biotechnology in Sweden include Alfa-Laval, KabiVitrum and Pharmacia AB. The latter company is a leader in separation technology for biological research and industry.

Other countries

Other countries with developing biotechnology companies include Italy, Israel, Australia, Brazil. Communist countries including the Soviet Union and China are also investing in the new technology.

Financing of biotechnology companies in Europe

In the United Kingdom there are a number of institutions investing in the biotechnology industry including Technical Development Capital, Prutech, Cogent and Advent. Technical Development Capital is the venture capital section of the Finance for Industry Group. It has investments in Imperial Biotechnology, Celltech and Plant Science. Prutech is a subsidiary of the insurance company Prudential and has investments in a number of British companies including Celltech. Cogent is a fund established by two insurance companies Commercial Union and Legal and General, while Advent is a venture fund established by Monsanto.

The venture capital system in Europe is relatively undeveloped and does not play a major role in financing biotechnology industry. Instead government programmes (as in Japan) are an important source of capital for the development of the industry and the provision of the necessary technological infrastructure.

The European Commission has also attempted to provide funding for biotechnology research and has a recently initiated a biotechnology Action Programme with an expenditure of 55m ECUs (European Units of Account, 1 ECU = $0.7) over five years. The European Commission argues that a concentrated effort is required in Europe if the continent is not to fall further behind the United States and Japan in the commercialization of biotechnology. However, the size of the research funding is too small to make a serious contribution to catching up with the United States or Japan while the continuing failure of Europe to develop continent-wide 'home' markets free from restrictive national barriers to trade is also a major problem in developing a European biotechnology industry. Europe will un-doubtedly contain some first-rate biotechnology companies but it cannot expect to be in front rank beside the United States and Japan if present trends continue.

4   Characteristics of the emerging industry

The biotechnology industry is emerging within a number of existing industries. As such biotechnological products will share the various characteristics of the relevant market with non-biotechnological products. The marketing of a biotechnology-based drug will obviously resemble the marketing of a conventional drug more than the marketing of biotechnology product in another industry.

The biotechnology industry can also be regarded as an emerging industry and thus having those features which characterize any emerging industry. Michael Porter has described some of the most important features of an emerging industry which include embryonic companies and spin-offs, early entry barriers, and technological uncertainty [104]. The way in which these factors apply to biotechnology is discussed below. It is useful, therefore, to view biotechnology both as a series of innovations within existing industry and also as an emerging industry itself and subject to the forces which characterize such an industry.

4.1 Characteristics of biotechnology markets

Pharmaceuticals

This industry is characterized by a globally integrated structure of manufacture, R & D, planning and distribution. It is dominated by American and European multinational enterprises. In industrialized countries an oligopolistic structure exists within the market for each therapeutic class and the overall success for inividual companies depends on having a few successful products in some market segments. Competition is on the basis of product differentiation, which in turn depends on R & D productivity. Quality of R & D is probably the key success factor in the industry. Price competition does not occur to any significant extent other than between drugs coming off patent and generic substitutes. This is because the main market for drugs is composed of doctors who prescribe on the basis of performance rather than price [110]. Pricing policies are also influenced by government regulation which can limit the rate of return on investment or the maximum permissible profits on different drugs. Patents play an important role in ensuring that companies receive adequate return on the massive R & D expenditures involved in developing and commercializing a new drug. Because of the extensive clinical testing required, a new drug takes from seven to ten years to develop in the United States and costs $70m. This acts as a significant entry barrier preventing smaller chemical companies from entering the ethical pharmaceutical business. Small companies can, however, manufacture generic substitutes (i.e., off-patent drugs) although this type of market entry is not relevant to biotechnology at this stage in its life cycle.

Biotechnology-based pharmaceuticals such as interferon, human growth hormone etc. will also be subject to the strict regulation that affects conventional pharmaceuticals and hence to large development costs. In cases where a biotechnology product is identical to an existing drug e.g., recombinant insulin it will nevertheless be treated as a 'new drug' under the regulations of the FDA.

In vitro diagnostics

In vitro diagnostics are also a part of the pharmaceutical industry but one which has important characteristics which distinguish it from drugs. At present a number of multinational enterprises and smaller national companies compete in a range of assays which are sold to hospitals, health centres, blood banks, etc. An increasing trend here is for diagnostics to be based at surgeries of general practitioners or in the home, and this trend combined with cost containment measures in national health care reimbursement policies will have significant effects on innovation in *in vitro* diagnostics.

Monoclonal antibody-based diagnostics represent a much less costly area than drugs. This is because the investment needed in R & D is much less and also because compliance with regulatory requirements in much less expensive. The manufacturers of Mab diagnostics have been able to show that their products are 'substantially equivalent' to previous assays and this has meant that such products could be marketed fairly rapidly.

Specialty chemicals

Specialty chemicals is a highly diversified industry with product areas as different as electronic chemicals, specialty polymers and biotechnology products. The end uses of biotechnology products vary widely depending on whether the product is used in food and feed industries or has other industrial applications.

Although some of the main product areas such as amino acids and enzymes are dominated by a small number of Japanese and European companies, small companies can compete successfully in the large number of niches available in the market for biotechnology-based specialty chemicals. Competition at present depends on price and biotechnology innovation in this industry is aiming initially to achieve economies in the production processes. At a later stage biotechnology-based specialty chemicals will be competing to an increased extend on product differentiation as companies develop new substances to compete functionally with existing chemicals. The specialty chemicals market is not subject to the same degree of regulation which occurs in pharmaceuticals. Foods and food additives are regulated by FDA. but other chemicals do not require FDA approval.

Food

Biotechnology innovation within the food industry is largely concerned with process innovations. It is possible to modify the microbial processes used in such industries as cheese making, brewing, baking, wine making etc. and new processes for existing foods can be developed through use of new specialty chemicals such as enzymes. This has not yet occurred to any significant extent. Competition here will also largely be on the basis of low cost processes.

Agricultural products

This includes animal health care and plant biotechnology. Animal health care, including animal drugs, vaccine and growth promoters is regarded as a sub-section of pharmaceuticals. The plant biotechnology business involves or will involve companies which sell live plants and/or seeds for agriculture, horticulture and the ornamental market. This type of activity varies greatly in the costs of R & D and market entry depending on the type of product and the genetic manipulation attempted. Micropropagation of commercial plants will be much less costly and involve a shorter developmental period than complex recombinant DNA work on such areas as drought resistance or herbicide resistance. Seed production is a local business due to climatic variation and it is therefore not possible to achieve global economies of scale in R & D as in pharmaceuticals.

Commodity chemicals

This industry consists of bulk organic and inorganic substances. Organic chemicals are obtained from petroleum and act as feedstocks for a wide range of other industries including rubber, plastics, polymers, synthetic fibres and pharmaceuticals, and fine chemicals. Commodity chemical plants typically cost hundreds of millions of dollars. The main entry barrier is not the cost of regulation but the cost of plant and associated distribution system. Competition is based on access to cheap raw materials and highly cost efficient processes.

The potential application of biotechnology here is to achieve cost reduction either by use of biological feedstock (biomass) and/or through replacement of chemical technology by microbial or enzymatic processes. Biotechnological processes cannot compete at present with commodity chemicals derived from petroleum and there is no indication that they will do so in the forseeable future. At present commodity chemicals are not a significant area for R & D investment in biotechnology.

4.2 The emerging industry

Biotechnology exhibits many of those characteristics which are found in any emerging industry. These include an early proliferation of small embryonic companies, typical early entry barriers and technological uncertainty which increases the risks of early participation in the industry and contributes to strategic uncertainty.

4.3 Embryonic companies

At the beginning of a new industry, risk is high and providing there is access to technology, entry is relatively easy. This leads to formation of a number of small entrepreneurial companies which may have spun off from larger companies or may be entirely new start-ups. These companies typically lead to further spin-offs as they grow and personnel leave to form their own companies.

This has been the experience in the biotechnology industry between the mid 1970s and the present. From 1976 to 1984 more than one hundred new biotechnology firms (NBFs) were formed in the United States. Genentech, one of the seminal NBFs was formed in that year (Cetus was already in existence since 1971) and in the subsequent three years 1977–1979, the number of NBFs formed in the United States was three, four and six respectively [99]. In 1980 this increased significantly to twenty-six and to forty-three in 1981. Since then the number of new start-ups has declined to twenty-two in 1982 and three in 1983. This decrease in the number of new start-ups reflects the increased difficulty experienced by NBFs in acquiring finance in addition to other entry barriers such as proprietary technology and the 'technological gap' between existing NBFs and potential new entrants.

4.4 Early entry barriers

Porter describes a number of entry barriers which characterize an industry at an early stage of its development. These include proprietary technology, access to distribution channels, access to raw materials and skilled labour, cost advantages due to experience, government regulation and risk which raises the effective opportunity cost of capital [104].

In biotechnology access to raw materials is not a significant factor, nor is access to skilled manpower (except in some countries). Cost advantage due to experience is also not relevant as the experience curve concept cannot be applied in the conventional sense to a rapidly changing technical environment. (See Chapter 9, Market Focus and Positioning). Capital barriers are closely related to government regulation since the regulatory class of a biotechnology product determines the cost of R & D necessary to develop and commercialize the product.

The main entry barriers in the biotechnology industry are therefore:

- proprietary technology both patented and secret;
- cost of R & D and regulation;
- access to distribution channels.

Proprietary technology

Product and process innovations in biotechnology are protected both by patent and by secrecy. There is no generic patent in the process of hybridoma production. In recombinant DNA, Stanford University has attempted to acquire a generic patent retrospectively, i.e., after the technique had diffused widely and had led to the formation of the NBFs. Stanford first filed the patent on behalf of Stanley Cohen and Herbert Boyer who claim to have invented recombinant DNA in the early 1970s. The process patent was granted in the United States in 1980 and the product patent in 1984 [20,62]. These patent applications were subject to considerable litigation.

Companies have been active in taking out new patents on individual products and processes many of which are also likely to be subject to litigation. For instance, Hoffmann-La Roche and Schering-Plough have been involved in legal action over patents on alpha interferon. There has been some confusion over whether patenting is worthwhile in the rapidly changing environment of the biotechnology industry. J. Leslie Glick [70], the Chairman of Genex believes that patenting is less important than secrecy in biotechnology while other leading companies such as Genentech and Cetus have pursued active patenting policies and have indicated that they intend to enforce these. In choosing whether or not to patent, a company would evaluate such factors as whether there is a likelihood of others discovering the invention independently, how easily the patent could be policed, the rate of technological change in the area, and whether or not the invention is likely to meet legal requirements for patenting [99]

Despite the confusion which has surrounded patents in biotechnology, they will probably play an important role in protecting proprietary information especially in pharmaceuticals where patenting has traditionally been important. In addition, regulatory agencies may move to extend patent life beyond the normal seventeen years to take into account the time spent in development. There is some evidence that recent biotechnology patents have already acted as entry barriers. The Japanese company Kyowa Hakko entered into an agreement with Genentech and Mitsubishi Chemical which allowed the two Japanese companies an exclusive licence for the Japanese market for tissue plasminogen activator (tPA). The Managing Director of Kyowa, Hirotoshi Samejima, is quoted as saying that Kyowa orginally tried to develop tissue plasminogen activator themselves but found that Genentech had the field sown up with a number of key patents [34].

Similarly companies wishing to enter various pharmaceutical markets at a later stage will find that all of the obvious routes using recombinant DNA are already proprietary. In diagnostics this is not likely to be such a problem because a wide range of technical solutions are available for a particular assay. Because scientific papers are produced by companies as well as patent documentation, knowledge of the molecular biology involved in the production of new biotechnology drugs will be widely available but it will also be unusable commercially.

Secrecy is also a very important method for protecting proprietary information within biotechnology companies and the policies adopted by American academics doing contract research for companies has led to the controversy that the biotechnology industry was inhibiting the free flow of information within the academic community. A company may choose to keep any or all of a particular process or product secret. In the case of substances made by recombinant DNA, the details of the expression system may be secret and also the details of the large-scale production, recovery, purification and quality control. The ability successfully to scale up and purify a product is often the key factor in a competitive R & D race with another company. Many companies have chosen to keep this type of bioengineering knowledge secret and diffusion of this knowledge will depend on personnel mobility within the industry. Such bioengineering expertise can take several years to acquire from start-up and the lack of access to this type of information is an important entry barrier.

Cost of R & D and regulation

The cost of entering the biotechnology industry for an NBF is related to the expenses involved in R & D. (For the established firm technology can be acquired by licensing or by company acquisition.) Research and development expenses are in turn related to the level of regulation so that the costs of entering any particular market are determined largely by the cost of regulation.

In the United States, the FDA, the Department of Agriculture (USDA) and the Environmental Protection Agency (EPA) all have significant regulatory authority. In Chapter 6 the issue of regulation is considered from the perspective of national competitiveness. Here it is relevant in relation to entry barriers.

The most highly regulated areas are human drugs, biologics and *in vivo* diagnostics. *In vivo* diagnostics involving the use of monoclonal antibodies inside the body will also be highly regulated. The substantial costs involved in meeting the regulations creates a barrier to the NBF which chooses to enter these markets. For those NBFs wishing to manufacture and market their own drugs, access to large amounts of capital over an extended period is necessary to finance the expensive clinical trials. Many NBFs may find it impossible to finance this process on their own and thus this will prevent them from entering the pharmaceutical market as independent manufacturers.

Other areas in health care are considerably less expensive than drugs. *In vitro* diagnostics do not require extensive clinical trials. In the United States manufacturers of Mab diagnostic kits are required to give the FDA ninety days notice before they can market a device and FDA determines during this period whether or not the device is 'substantially equivalent' to previous devices. The cost of supplying monoclonal antibodies to manufacturers of *in vitro* diagnostic kits has been estimated to be $3.5m to $4m over a three-year period in the United States, while developing the final diagnostic kit may cost between five and ten times more [111].

In specialty chemicals the degree of regulation will depend on the end use. Food and feed ingredients such as flavours, amino acids, etc. will be regulated by the FDA and will require premarketing approval. However, specialty chemicals produced for various industrial applications will not require regulatory approval. The cost of regulation will vary considerably across different groups of specialty chemicals.

For animal health care products such as drugs, vaccines and diagnostics the cost of R & D will be similar to human products but regulatory requirements are less strict so that products can be got to market faster and with less overall expense.

Although the cost of regulation is the main factor affecting the commercialization of different groups of products it is not the only one. The cost of R & D itself varies. Some types of research are inherently more expensive than others because of the size of the research teams required, the inherent technical difficulty and the differing requirements for investment in research facilities and pilot plant. The production of *in vitro* diagnostic kits does not involve investment in sterile fermentation plant which is required for recombinant DNA pharmaceuticals and this can allow a small company to enter diagnostics more easily than the pharmaceuticals. Some potential application areas of biotechnology such as single cell protein or commodity chemicals require massive investment in plant and this has been one of the factors contributing to the lack of investment in these areas in comparison to health care and agriculture.

Access to distribution channels

This is a major entry barrier facing NBFs which in general do not manufacture and distribute their own products. The NBFs have been forced by their lack of a distribution system to have their products manufactured and/or distributed by established firms. This has led to a wide range of marketing agreements between NBFs and established firms. The nature of the distribution channels depends on the application area of biotechnology and on the country in question. Thus in the United Kingdom pharmaceutical industry there is little variation among manufacturers in the channels of distribution used [110]. In the United States, however, the selection of the appropriate channels of distribution is crucial to marketing success. While wholesalers are of central importance in distribution to retail chemists and hospitals in the United Kingdom, direct selling to large retail pharmacists is essential in the United States.

New biotechnology firms developing biotechnology-based pharmaceuticals are not in a position to engage in the extensive promotion or marketing of new drugs that involves a sizeable sales force. In addition, many of the drugs which these companies are developing are likely to be prescribed by specialists rather than general practitioners. Therefore, the normal distribution channels used in the pharmaceutical industry are not accessible to NBFs. NBFs have faced this problem in three main ways:

- by licensing their technology to established firms which do have access to distribution channels;
- by in-house manufacture combined with marketing agreements with companies in the home country and overseas;

● by marketing of their drugs to a select group of specialists.

Examples of licensing include Genentech's agreement with Eli-Lilly, on human insulin and Biogen's agreement with Schering-Plough on alpha interferon. Under these agreements manufacturing and marketing are performed by the major established firm.

In the process of forward integration NBFs engaged in pharmaceuticals will still need to distribute their products internationally and this involves a marketing arrangement with a foreign partner until such time as a foreign subsidiary is established (cf. Chapter 9). To overcome the limitations of their size, NBFs can also attempt marketing new biotechnology-based drugs to a small number of specialists who can be reached with a small sales force. An example here is Genentech which plans to market its human growth hormone to a small number of hospital-based endocrinologists in the United States and to rely on foreign partners for overseas distribution.

In the area of specialty chemicals, access to distribution channels may be considerably easier than for pharmaceuticals. A small number of companies may hold a large share of the domestic market for say an industrial enzyme and NBF can thus distribute its product directly to these end users without a marketing team.

4.5 Technological uncertainty

The early stages of a new industry are typically characterized by a high degree of technological uncertainty. This has a number of effects. It raises the opportunity cost of capital because it increases the factor of risk. It leads to an initial reluctance to invest by established companies affected by the new technology and this behaviour can be seen in the activities of the established pharmaceutical companies which in general did not invest until after 1980. It also raises major strategic questions for the firm as we shall see below (Chapters 8 and 9). Technological uncertainty in the emerging biotechnology industry exists at both product and process levels.

Process uncertainty

Process uncertainty arises from, the simultaneous emergence of a number of alternative new processes and doubt about their advantages and disadvantages relative to existing processes.

Table 4.1   Competing technologies: criteria for the choice of bacteria, yeast or fungi as suitable hosts for recombinant DNA

	Bacteria	Yeast	Strepto-mycetes	Mammalian cells
Ease of gene manipulation	+++	++	+	++
Development of expression systems	+++	++	+	++
High density growth	+++	+++	+++	+ (micro-carriers)
Ease of cell harvesting	++	++	+++	+
Ability to process cell material	+++	++	+	++
Excretion of required products	±	++	++	++
Glycosylation of required products	−	+	±	++

Source: Dr. A. Pickett. Porton International Ltd. Reproduced by permission.

Recombinant DNA technology uses a variety of hosts (Table 4.1) including yeast, fungi, mammalian cells and the bacteria *E.coli* or *B. subtilis.* Depending on whether mammalian cells or micro-organisms are chosen as host, the final product may be glycosylated or unglycosylated (with or without a sugar group) and this may have an impact on the performance of the product clinically. This choice of production methodology may have crucial effects on the efficacy of some pharmaceutical proteins. The choice of host and expression system can also have significant effects on the process economics depending on such factors as whether production of the protein is extracellular or intracellular; that is, whether it is exported by the cells into the medium or retained within the cells and concentrated thus requiring cell lysis before extraction and purification. A calculation carried out by Genex Corporation compares the economics of extracellular vs. intracellular production (Table 4.2). The recovery efficency for extracellular production is 80 per cent as against 50 per cent for intracellular. Raw materials and labour costs

Table 4.2 Process Economics: Intracellular production of a specific protein vs. export of that protein in manufacturing facilities of identical space.

	Production costs ($m)	
	Intracellular protein (5% of total protein)	*Extracellular protein (10 g/l)*
Raw materials	1.60	0.17
Labour	2.30	0.42
Utilities	0.25	0.14
Equipment	1.14	1.11
Buildings	1.00	1.00
Direct costs	6.29	2.84
Overhead (60% of labour)	1.38	0.25
Annual production (lb)	10,000	25,000
Unit cost ($/lb)	767	124

Source: [70] Paper submitted to Institute of Health Economics & Technology Assessment. 27 November. 1981. Reproduced by permission

are significantly greater for intracellular production and this results in a unit cost of $767 per lb for intracellular production as against $124 per lb for extracellular production (1981).

Product uncertainty

Uncertainty also exists over which product to pick for the new biotechnology markets. For instance, in cancer research a wide variety of new products and technologies are simultaneously under development (Table 4.3). Companies cannot devote equal resources to all of these products and must choose one or a few in an environment where clinical data is incomplete. Similarly, in diagnostics there are a range of different technologies which can be used for any particular assay (Table 4.4). A further level of product uncertainty is introduced by the development of second generation products and this is related to the choice of technological strategy which is discussed in Chapter 9.

Table 4.3   Alternative new products/technologies for cancer therapy

• Alpha interferon	• Macrophage activating factor
• Gamma interferon	• Tumour necrosis factor
• Beta interferon	• Lymphotoxin
• Interleukin-2	• Immunotoxins
• B cell growth factors	• Applications of oncogene
• B cell differentiation factors	research

Table 4.4   Choice of diagnostic technologies for an assay

- Radioimmunoassay — polyclonal
- Radioimmunoassay — monoclonal
- Enzyme-linked immunoassay — monoclonal
- Ensyme-linked immunoassay — polyclonal
- Chemiluminescence
- DNA probe
- New enzyme-linked immunoassay technologies

5 The role of science

The biotechnology industry is generally regarded as a high-technology industry analogous to information technology. This concept of high technology implies knowledge intensity. In biotechnology, this knowledge intensity involves a very intimate relationship between basic science and commercial activity and this relationship has profound implications for the structure of the developing industry. Before exploring this relationship and its consequences, it is useful to examine the origins of the biotechnology industry and to contrast this with the origins of the semiconductor industry.

5.1 Origins of biotechnology and comparison with semiconductors

Biotechnology processes such as fermentation were well established after the Second World War and fermentation was used to produce a variety of new products such as steroids, antibiotics and fine chemicals. The 'new' biotechnology industry with which we are concerned here arises from a wave of innovation occurring within this conventional fermentation-based industry and results from the application of a number of new techniques involving, for example, monoclonal antibodies, genetic engineering, cell culture, enzyme immobilization and protein engineering. It is not possible to consider the origins of all of these technologies here so we will examine two of the most important namely, genetic engineering and monoclonal antibodies.

The molecular biologist Maxine Singer* has described the main strands of research which led up to the ability to perform genetic engineering. This resulted from the coming together of two major lines of molecular biology research, DNA and enzyme studies over a period of the last thirty-five years. The 'DNA Revolution' involved

---

* Setlow, J.K., Hollander, A., 1971. (eds.) *Genetic engineering: principles and methods*, Vol. 1, p. 113. *Plenum Press*, New York.

the identification of DNA as the genetic material and the pioneering work on its structure and that of the genetic code. The equally important 'enzyme revolution', however, was essential for the development of the ability to manipulate genes. Over the same period, much information was obtained on the role of these important macromolecules in catalysing reactions in living systems. It was the study of a curious phenomenon called 'host-controlled modification and restriction' which led to the identification of one of the central tools of recombinant DNA, restriction enzymes. Researchers studying viruses which grow in bacteria (phages) found that the ability of the bacteriophage to reproduce in a particular cell type depends on the cell type in which the phage was previously grown. This phenomenon could not be understood in terms of the genetics of that time. Subsequent workers such as Werner Arber showed that the phenomenon of restriction involved DNA breakdown and Arber argued that this breakdown probably involved a highly specific initial cleavage and was related to the specificity of the DNA sequence (although there was no information supporting this at that time). Arber and co-workers described (in the mid to late 1960s) the biochemical basis of restriction. This was caused by restriction enzymes which cut the DNA. Early studies concentrated on Type I restriction enzymes which do not cut DNA at specific base sequences. In 1970 Smith with Wilcox and with Kelly described Type II restriction enzymes which cut DNA at specific sequences. This was the critical event in the development of recombinant DNA technology as it provided molecular biologists with the tools to cut DNA accurately at desired locations and so isolate genes. The development of recombinant DNA research also depended, however, on the discovery of other enzymes with other functions such as DNA ligase which joins strands of DNA and the enzyme reverse transcriptase which allowed the development of the subsequent ability to clone mammalian genes in microorganisms.

Two pioneers of recombinant DNA research were Stanley Cohen of Stanford and Herbert Boyer of the University of San Francisco. They applied for a patent on the technique in 1974. There was much early confusion over the patent and throughout the 1970s the technique diffused widely throughout the scientific community and led to major advances in molecular biology

The commercial significance was, however, soon appreciated and in 1976 Boyer teamed up with industrialist Robert Swanson to form Genentech to exploit the technology commercially. Another company

Cetus (which was in existence five years previously) soon realized the potential of recombinant DNA and began to invest in R & D. In the late 1970s more new companies were formed (often through academics leaving university) and by 1981 over one hundred new companies had been started with many of them concentrating on genetic engineering applications.

The technique of making Mabs was discovered by Cesar Milstein and George Köhler in 1975 at the Laboratory of Molecular Biology in Cambridge, England. At the time Köhler, then a post-doctoral scientist was studying immunogloblin genes at the laboratory of Cesar Milstein. The two scientists had the idea, of fusing two different types of cell, a lymphocyte to produce a specific type of antibody and a cancer cell to 'immortalize' the resulting cell. Their paper describing the discovery was reported in *Nature* in May 1975. The possible application of their discovery was evident to them and before publication Milstein got in touch with a British Government official to suggest that the technique should be patented. According to Köhler 'Cesar said there was no response, so we published'.* As a result of this the British Government lost the opportunity to patent one of the key techniques of the new biotechnology. Subsequently the technique was widely adopted by researchers around the world and was applied in companies such as Centocor, Hybritech, etc.

Biotechnology emerged, therefore, as a direct result of technology transfer from the universities which originally were the only places where the expertise existed. In considering, the emergence of biotechnology and the relationship of the 'science' to the 'technology' it is useful to compare biotechnology to semiconductors.

The semiconductor industry originated from the discovery of the transistor at Bell Labs in 1948 by Shockley [56,91]. Bell Labs was the research arm of AT & T. Bell Labs spent considerable amounts (for the time) on transistor and semiconductor research, $57m up to the end of 1964 [56].

The Bell basic patents in transistors were made available to all comers on payment of $25,000 dollars advance royalty, partly as a result of an anti-trust suit filed by the Department of Justice in 1949 and finally settled in 1956 [99]. Bell Labs resulted in many spin-off companies which pioneered major innovations such as the integrated circuit and these in turn led to formation of other companies. There

---

* *Science*, 1982. Hybridomas: the making of a revolution. 26 February, Vol. 215, p. 1074.

are considerable similarities and dissimilarities between the emergence of these two industries and these include:

- the nature of university/industry linkages;
- the patent situation;
- the relation of product to process innovation;
- the role of government.

University/industry linkages

Both semiconductors and biotechnology involved close interaction between university-based expertise and the companies. In the case of semiconductors, however, federal funds were provided to universities to help reduce the gap between the technology levels of the universities and that of the companies. The American semiconductor industry became concentrated around the major recipients of federal funding, San Francisco and Boston. The closeness of the universities and the semiconductor firms facilitated links between them.

In biotechnology, the universities played a crucial role in transferring technology to the new companies. Many of the company personnel, both R & D and managerial, came from the universities. An important difference therefore, was that in semiconductors the original discovery and early expertise were industry based with university/industry interaction emerging later, while the biotechnology industry originated by the transfer of the original university-based expertise into new and established companies. As Ellis has pointed out, the NBFs can be regarded as university spin-offs in an analogous manner to conventional company spin-offs [38].

Patent situation

In semiconductors the original transistor patent was made widely available on payment of a royalty fee. Subsequently, the patenting of integrated circuits has not played a major role in the development of the industry due partly to the ease of copying integrated circuits. In the pharmaceutical industry on the other hand, long lead times and massive R & D budgets can only be recouped under patent protection. Hence the products of biotechnology in pharmaceuticals and diagnostics are being patented and patents will continue to play an important role. At the emergence of the biotechnology industry, there were no patents to prevent diffusion of the technology. The

Stanford patent, applied for in 1974 and granted in 1980 was essentially an attempt to gain proprietary rights to the technology *after* it had become diffused. This rapid diffusion occurred because biotechnology emerged in a university setting with free publication of the research results. In the case of monoclonals, no patents of a genetic nature were taken out, although many companies are now applying for patents on specific products.

Product and process innovation

Some analysts of the semiconductor industry argue that there is little distinction between product and process innovation in the industry. The 'technological life cycle' theory suggests that in the early stages of a new industry product innovation predominates and that as the industry matures, this slows down while process innovations increase. However, in semiconductors "process innovations far from being a minor factor in the early years of the industry has always been crucial to commercially successful product innovations" (Mowery) [91]. This is also the case in biotechnology. A new product based on genetic engineering such as alpha-interferon cannot be considered separately from the process innovation which allows its commercialization. In addition there is a very special relationship between basic and applied research in biotechnology: this is explored more fully below.

The role of government

In the semiconductor industry, military procurement and the standards laid down for the quality of components played a major role in assisting early industry development [91]. United States government policy towards semiconductors was also a significant factor. The Department of Defense and NASA were major sources of R & D funding within industry. The federal government was also involved in demonstration projects and in assistance for development of production technology.

There is no parallel in the case of the development of biotechnology within the United States. Here, federal involvement has been very limited, although the basic research programmes which led to the development of biotechnology were federally funded. The major role played by the United States government has been in the development of regulations for R & D and manufacturing safety through agencies such as the National Institutes of Health and the FDA. Outside the United States, however, national governments have 'targeted' biotechnology as an important priority for development. This is especially important in countries such as Japan, France and the United Kingdom. Such governmental actions are likely to contribute both to the competitiveness of the firm and national competitiveness: they are explored more fully below in Chapter 6 on 'The role of governments'.

5.2 Basic and applied science in the new companies

A special relationship between basic and applied research (or science and technology) has developed within the new companies. Characteristically, this makes it difficult to distinguish basic from applied research and involves concentration of scientific talent and performance within commercially orientated organizations. Before discussing this it is necessary to look at the definitions of basic and applied research. Science is often regarded as a body of knowledge of the natural world; technology as the knowledge and artefacts involved in industrial processes. The purpose of basic research is the advancement of fundamental knowledge while that of applied research is the application of that knowledge in the creation of technology. Many authors have previously considered the definitions of science and technology. Thus Price [107] draws a sharp distinction between science which leads to the output of papers and technology which does not. Schmookler [108], however, considers technology to include published papers as well as other types of information, and defines technology as a 'pool of knowledge of the industrial arts'. In general we can identify some general characteristics of basic and applied research as shown in Table 5.1.

The new biotechnology firms, most of which were founded by scientists are research-intensive companies as discussed above. The environment within the companies, with its scientific curiosity and freedom, resembles in many ways that within university departments and this has greatly contributed to the ability of the companies to attract and hold first-class scientific expertise. (Compare Chapter 9 'Innovative culture'.)

The application of molecular biology to new products (especially in health care) is crucial to most of these companies and to understand the relationship between the science and the technology

Table 5.1 Characteristics of basic and applied research

Basic research	Applied research
Intention is new knowledge	Intention is new technology or application
Open-ended	Directed
Long time frame	Short–medium time frame
Freely published	Generally proprietary or restricted publication
Carried out in universities	Carried out in companies

it is necessary to consider the subject of molecular biology. Modern molecular biology which is only thirty years old is probably the fastest growing area of knowledge in the world today and there is no sign that the exponential growth is decreasing. The scientists who staff the new companies were often participating in this phenomenal increase in knowledge before joining the company and continue to do so within the corporate structure. Companies with a high concentration of scientific talent are generating much new knowledge as well as applying existing knowledge (Table 5.3). Genentech, for instance, has published over one hundred scientific papers. It is possible to construct a spectrum from molecular biology of a purely basic nature to the interaction of molecular biology with other disciplines as in areas of applied research. Between those investigations which are basic and have no immediately identifiable application or relevance and technology development is a 'grey area' of research in which fundamental discoveries are immediately relevant to technology and are pursued with that end in mind.

Table 5.2 Spectrum of basic to applied research in biotechnology

Basic	Applications-oriented science	Applied
Cell structure	Gene cloning	Optimisation of recombinant DNA processes and scale-up
Molecular structure and function	Gene structure	
	Cancer	
	Immune system	

The new companies are carrying out research into areas such as cancer, the immune system, new gene structure, new protein structure, etc. In cancer, for instance, the companies have played an important role in the development of knowledge about substances such as interferon, interleukin 2, and immunotoxins. The discovery of a novel substance, with say anti-cancer effects or a new gene implicated in cancer, is a contribution to basic knowledge. But since the intention and immediate use is commercial, it is also a technological innovation. Such discoveries are simultaneously both science and technology or rather we can make no meaningful distinctions between basic and applied research and its objectives. This form of research is termed here applications-oriented science. This relationship between basic and applied research and also the particular circumstances under which the new industry arose have a number of important implications.

5.3 Implications

The special relationship between science and technology in biotechnology has a number of important implications for industry structure and competitive strategy. These are:

- a capability in basic research is essential for a firm's competitive ability;
- rapid generation of new proprietary knowledge presents a significant entry barrier;
- the industry is emerging globally.

Basic research and competitiveness

There has been much previous debate on the role of basic research within the corporation. Freeman, writing in 1974 [56], stated that

it is fashionable to deride in-house fundamental research and to regard it as an expensive toy or a white elephant.

From his own research and that of others, Freeman concludes that fundamental research, whilst not essential to an offensive innovation strategy, is often a valuable means of access to new and old knowledge generated outside the firm, as well as a source of new ideas within the firm.

Undoubtedly, this is true in many industries in which companies can access external scientific knowledge and use it to generate technical innovations within the firm. However, in the biotechnology industry, firms must participate in 'applications-oriented science' which implies basic research. In many of the leading applications of biotechnology and especially in health care, the nature of the R & D required implies also a strong capability to generate new basic knowledge. There is, therefore, no conflict between basic and applied research and we should expect that the firm which is highly productive in its applied R & D will simultaneously be productive in its basic research achievements. In this way an activity, namely basic research, which was previously regarded as being the province of non-profit making organizations now moves into the commercial arena. Science becomes a competitive activity, not in the old sense of reputation, but in a commercial sense also and firms need to attract and hold the best scientific talent they can get. In this sense biotechnology represents not only the application of biology to industry but the commercialization of biology itself.

Rapid generation of proprietary research

As an industry develops there is a diffusion of industrial know-how. However, the rate at which diffusion occurs depends on the rate at which new proprietary information is produced. If there is a very rapid rate of generation of proprietary knowledge then this will tend to offset technological diffusion.

In biotechnology, new knowledge is both rapidly developed and highly protected through a combination of patents and secrecy. This creates a significant entry barrier. The fact that firms need a capability in 'application-oriented science' as well as access to production technology creates a further barrier. This barrier can only be overcome with large financial resources and so it is less of a barrier to the established firm than to a NBF.

Table 5.3 Selected scientific achievements of new biotechnology firms in molecular biology 1984

Genentech	• Cloning and expression of lymphotoxin gene
	• Cloning of gene for tumour necrosis factor
	• Cloning and expression of human oncogene
	• Partial characterization of human tissue plasminogen activator
Biotechnica International	• Development of new high-level expression system
Immunex Corp.	• Cloning of interleukin-2 receptor gene
Celltech	• Purification of recombinant prochymosin
Centocor	• Cloning and expression of genes from human T-cell leukaemia virus III
Cetus Corp.	• New techniques for synthetic vaccines including feline leukaemia virus

Source: Scientific papers including [2, 69, 71, 85, 88]

Global emergence

Because knowledge of basic molecular biology techniques has diffused widely, the biotechnology industry is growing in all parts of the world. This results from the free publication of much seminal molecular biology and the fact that research teams were pursuing similar lines of inquiry in different countries. Biotechnology is, of course, more than molecular biology and bioprocessing capabilities were already global in extent before the first NBFs were set up. Some existing bioprocessing industries were able to adapt to the new technology. This contrasts with the situation in semiconductors, where the industry emerged first in one country (the United States) before spreading around the world. The global emergence of the biotechnology industry implies that a firm may face competition in new product development from widely diverse countries and it, therefore, requires a global perspective in its operations. This is explored more fully in Chapter 9.

6 The role of governments

Government influence on the competitiveness of national biotechnology companies can be substantial. Government intervention may consist both of new actions and initiatives and of the effect of existing laws and regulations and it may have the effect of increasing or decreasing competitiveness. Government policies for biotechnology are designed obviously to increase national competitiveness and to modify prior legislation or regulation which is inhibiting its development.

The role of governments in the biotechnology industry must be seen in the context of national innovation policies. During the 1970s a number of influential OECD reports highlighted the role of governments in encouraging the development of high technology industries. The increased competition which Western countries experienced in that decade from the newly industrializing countries also had the effect of increasing awareness within the leading industrial nations of the need to develop knowledge-intensive industries. This trend was greatly aided by developments in the world electronics industry. The targeting of this industry and especially of the area of components by the Japanese received considerable publicity. The Japanese Very Large Scale Integration Project which involved a government/industry partnership acted as the prototype for governmental intervention aimed to achieve parity or superiority with overseas competitor nations [73]. Since then a number of Western nations have adopted long-term development plans for their electronics industries and the European Commission sponsors the ESPRIT programme designed to promote European competitiveness in this technology. When 'biotechnology' first received major publicity in the late 1970s the groundwork had already been laid for an active governmental role.

This chapter discusses the forms of government intervention and the issues involved and then considers the effectiveness of these measures in contributing to national competitiveness.

6.1 Targeting policies

Targeting means the identification by government (or its agencies) of an important national priority and the development of an active state role in its development. Such actions can include the development of infrastructure, the adjustment of tax, patent, export and regulatory laws, direct investment in industry, provision of venture capital, and the encouragement of industry consortia to maximize economies of scale. Countries pursuing active targeting policies in biotechnology include France, Britain and Japan. In the United Kingdom, a government report (The Spinks Report) made wide-ranging recommendations in 1980, many of which were subsequently acted on. The government, through the British Technology Group took a minority interest in the NBF Celltech which was established in late 1980 and is now one of the leading companies involved in the commercialization of monoclonal antibodies.

The Department of Trade and Industry (DTI) has been given the job of co-ordinating activities and is spending £16m over three years from 1982 to 1985 on assistance to centres of excellence, demonstration projects and other programmes. The DTI is also responsible for establishing with the Atomic Energy Research Establishment (AERE), Harwell, and Warren Spring Laboratory, a "downstream processing club", (BIOSEP) which is an industrial research consortium [113].

In France the Government has developed a 'Mobilisation Programme' for biotechnology as part of a general programme in high technology. In Japan the Ministry for International Trade and Industry (MITI) has a ten-year programme with fourteen companies designed to develop advanced biotechnology processes with a total cost of about $100m [99].

The targeting policies of those countries can be distinguished on the basis of the type of industry that they are aimed at. Thus in Japan, entrepreneurial high technology start-ups will not play any significant role in the development of biotechnology within that country. Instead large established companies in the chemical, pharmaceutical, energy and food sectors will be involved. In the United Kingdom established chemical and pharmaceutical companies are regarded as playing an important role but emphasis is also placed on the American-style creation and fostering of NBFs.

The United States does not pursue a targeting policy, although its governmental agencies are very involved in regulatory affairs as

Table 6.1 Typology of governmental policies for biotechnology

Innovation policy type	Established firms predominate	NBFs also important
'Climate only'		United States
'Targeting'	France West Germany Japan	United Kingdom

described below. It relies on its existing tax laws, generous funding for basic research and an entrepreneurial climate for the development of the industry and this has so far been highly successful. Recently, there have been some moves towards suggesting a more interventionist role for the government. The report of the Department of Commerce *High Technology Industries: Profiles and Outlooks, Biotechnology* has made a range of such recommendations. However, there is no evidence to date that the United States government is likely to take on a more active role.

National biotechnology policies are of two broad types; those which pursue active targeting and those which rely on climate measures. It is also possible to distinguish between countries where new technology-intensive firms are playing an important role and those in which biotechnology developments can be expected only or largely from established firms (Table 6.1).

6.2 Regulation

The major regulations affecting the development of the biotechnology industry are those concerned with product approval and to a lesser extent export of drugs. (The regulation of recombinant DNA research was an issue in the 1970s.)

In the United States, products are regulated by the Food and Drugs Administration (FDA), the Department of Agriculture (USDA) and Environmental Protection Agency (EPA). The FDA regulates food and drugs according to a series of classes: human drugs, human biologics (vaccine, serum, toxin), medical devices, food and food ingredients, animal feeds, feed additives and devices and veterinary medicines. As most of the substances produced by biotechnology companies fall into these classes. FDA regulation is of major importance. New drugs such as interferon produced through recombinant DNA have to undergo the extensive set of animal and human clinical trials required for all other drugs. The total length of time involved can be up to seven years and R & D costs per new drug can be as high as $70m. This applies even if drugs made by recombinant DNA are identical to existing drugs. Diagnostic devices incorporating monoclonal antibodies are not as heavily regulated as new drugs. The USDA has authority in the area of animal biologics and premarket clearance is necessary. Some duplication and inter-agency conflict has arisen here as both FDA and USDA have indicated that they will regulate veterinary pharmaceuticals. This could lead to these products being subject to double regulation. The Environmental Protection Agency (EPA) has wide-ranging powers in the areas of insecticides and pesticides. The EPA issued guidelines in late 1984 on how it will regulate biotechnology products.

In Europe, the product approval process, although substantially similar to that in the United States, results in products such as pharmaceuticals being brought to market more rapidly. The American system results in substantial delays and costs in obtaining premarket approval. In contrast, the British system will allow a speedier access to the market. In the United Kingdom new biotechnology-produced drugs which are identical to existing drugs require product and manufacturing licences but they do not require the complete documentation that would be required for a new drug. In Switzerland, the drug approval process takes only six to ten months. Regulatory approval is not required before clinical trials are begun. Also less documentation is required for drugs that are not entirely 'new'. Thus drugs manufactured through recombinant DNA technology will suffer less delay than in the US.

In Japan, the product approval process for new drugs is similar to that in the United States in that extensive delays are involved due to the involvement of a number of agencies.

Other important issues involving governments are the questions of non-tariff trade barriers, market access and export regulations. Non-tariff trade barriers include health and safety standards and certification systems, subsidies and price regulation. In the pharma-ceutical industry at present non-tariff trade barriers and market access are already major problems. For example, American pharma-ceutical exports to Japan have been faced with a discriminatory certification system. Some progress has been made in dismantling this in recent Japanese legislation but barriers still remain. As the biotechnology industry develops world-wide, it is expected that governments will use this type of barrier to trade to protect their own industries.

American regulations concerning the exports of drugs are having an impact on biotechnology products. The FDA prohibits the export of unapproved new drugs except in cases where foreign governments specifically request it. This applies even in cases where the drug is approved in the importing country. This has implications for the American pharmaceutical and biotechnology industries which have been highlighted in two recent reports, that of the OTA [99], and that of the Department of Commmerce [113]. Both of these reports argue that this is likely to lead to transfer of American technology and jobs abroad to countries 'where the law permits the export of unapproved drugs' [OTA, 99]. In addition, the American NBFs which are involved in licensing agreements and joint ventures with foreign firms may have to transfer the new technology abroad if they are not permitted to supply bulk ingredients. Genentech has stated to the Department of Commerce [113] that this regulation adversely affects its business as it is unable to supply pharmaceutical markets in Europe, Canada and Japan from its new and costly production facility and it proposes existing regulations should be amended to permit export of new unapproved drugs principally to developed countries.

The American government export controls imposed for security reasons also affect the exports of biotechnology companies. The Export Administration Act of 1979 established export controls on biotechnology to September 1983. A new act is now in preparation. Multilateral export control measures are administered through COCOM, representing NATO countries plus Japan and Portugal. Biotechnology products and processes, including micro-organisms, plasmids, vectors, and equipment such as large-scale cell culture, affinity purification technology, filters and automated control systems may have potential biological warfare uses in countries hostile to the West. Although COCOM does not have any current export controls in the biotechnology area, it is likely that this issue will be considered by COCOM in the near future, as it is becoming evident that unilateral American export controls will not be effective in stemming the flow of technology to Eastern bloc countries.

In summary, there are a range of issues concerned with the regulation and export of biotechnology products and processes. The speed with which a product can be introduced to market and the conditions governing its export will have important implications for the competitiveness of a firm pursuing such a product. Governmental action in the area of regulation can improve the competitiveness of companies. Such actions can include the following:

- increase the speed of the product approval process by instituting abbreviated data requirements for biotechnology-derived drugs;
- devise a more streamlined regulatory process through rationalization of agency functions and elimination of duplication;
- amend export regulations to facilitate the export of biotechnology products by national companies;
- engage in bilateral negotiations with other governments to improve market access for national companies.

6.3   R & D policy and infrastructure

In all of the countries involved in biotechnology, governments fund both basic and applied research and also assist in building up the technical infrastructure.

In the United States basic research at universities is funded through governmental agencies such as the National Institutes of Health (NIH), National Science Foundation (NSF) and government Departments such as Agriculture (USDA) and Energy (DOE). In the United Kingdom, research councils such as the Medical Research Council (MRC) and Science and Engineering Research Council (SERC) fund research.

Applied research and co-operative university/industry research is also funded in a number of countries. In Japan more governmental funding is allocated to applied research (unlike the United States, where most of the money goes into basic research). The Japanese Ministry for International Trade & Industry (MITI) has organized fourteen companies into a research consortium and is allocating $43m to bioreacter projects, $43m to recombinant DNA projects and $17 to $22m to the mass cell culture projects over ten years. In the United States, the Small Business Innovation Research Programme (SBIR) of the National Science Foundation supported applied research in industry in the 1970s [99]. In the early 1980s it was proposed that an expanded SBIR programme be introduced. In July 1982, the Small Business Innovation Act was signed into law by the

President. The SBIR law requires each federal agency to set aside a certain percentage of its R & D budget for SBIR programmes. The scheme is in three phases. In Phase I a preliminary grant of up to $50,000 is made to the firm. Phase II which is the main research phase, provides grants of up to $500,000. In Phase III, federal support ends and follow-on financing of the R & D must be obtained from venture capital sources. The programme thereby links governmental support for R & D to the venture capital process [99].

In the United Kingdom, a particularly innovative programme of joint research is the 'downstream processing club'. This 'club' or research consortium involves two research institutes, AERE Harwell and Warren Spring Laboratory, in association with a number of British companies. The objective is to carry out research into improved separation and purification of products from bioreactors. The consortium was initially established using governmental funds from the Department of Trade and Industry. Members of the 'club' will have access to the results of the research. Another British research consortium involves the Leicester Biocentre at Leicester University which will carry out fundamental research in molecular genetics. The centre is being sponsored, in part, by five companies, Dalgety – Spillers, Distillers, Gallagher, John Brown and Whitbread. The industrial sponsors and the university will jointly own any results emerging from the research.

As well as assisting innovation within firms and contributing to the establishment of 'centres of excellence', state agencies are also involved in providing adequate levels of trained manpower. Access to specialized expertise is one of the major features which assist national competitiveness. Consequently, most of the countries involved in biotechnology have taken steps to improve their education and training facilities.

Government action in the area of R & D funding and infrastructural development can have significant impact on the development of any new industry. This has been shown by the development of the American semiconductor industry where federal government support played an important role in developing expertise within the universities. In biotechnology, governments can increase national competitive ability by a range of measures as follows:

- identify basic and applied research priorities and fund accordingly;
- initiate programmes for joint university/industry research;
- sponsor industrial research consortia;

- fund the establishment of university or non-university based centres of excellence.
- identify critical manpower shortages and take remedial action.

6.4 Tax, financial incentives and climate

Tax and financial incentives have been identified by governmental reports as major factors contributing to the successful development of a biotechnology industry. These measures belong to the 'climate' type of innovation policy, in that they are non-specific and apply to all or nearly all branches of industry. Incentives can be created not only by policy on company taxation, capital formation, and R & D spending but also by the financial climate for enterprise and the development of the venture capital system.

Established companies can finance biotechnology R & D investments through use of their revenue or debt and these companies can benefit from the various tax provisions for innovative activity in their countries. A summary of how innovative activities are treated in various countries is reproduced below (Table 6.2).

For the NBFs the availability of capital to fund R & D, scale-up and clinical trials is very important. In European countries there is significantly less venture capital available than in the United States and this is one of the reasons why the number of NBFs in European countries is relatively low. The American financing mechanism of R & D limited partnerships is an important source of funding for NBFs and allows American NBFs to fund costly clinical trials in a manner not open to British or Japanese companies.

6.5 Anti-trust environment

The United States and other countries have anti-trust laws designed to encourage maximum competition and discourage cartels. This issue is more relevant to the US than to other countries due to the substantial penalties and the way in which the law is administered. There is some concern that anti-trust laws may be applicable to R & D joint ventures and licensing agreements in the biotechnology industry. A document produced by the United States Department of Justice in 1980, *Anti-Trust Guide Concerning Research Joint Ventures* provides some guidelines on the legality of these joint ventures [99]. This depends on:

Table 6.2 Comparative tax treatment of innovation activities

Country	Capital expenditure on R & D	Venture capital investments in new firms
United States	As for other depreciable assets	Research and development limited partnerships pooling of investment funds in investment companies
Japan	100% depreciation allowance for member firms of research association	No special provisions
Federal Republic of Germany	Depreciated as for other assets	No special provisions
United Kingdom	100% tax allowance for research assets Allowances for both capital and current expenditure	No special provisions
France	50% of cost depreciable in first year with balance depreciable over useful life	Businesses which purchase shares in Qualified Research Companies and shares in Innovation Finance Companies may deduct 50% of the cost of the shares in the year of acquisition

Source: Compiled from Table 53 of OTA Report [99]

- the percentage of the industry participating as shareholders;
- the identity of the shareholders;
- whether risks and costs justify a joint project;
- whether the research is closer to basic or applied (the further removed it is from market effects the more likely it is to be acceptable);
- the number of actual and potential competitors in the industry;
- how narrow the joint field of activity will be.

Recently there have been some attempts in the United States to alter the existing legislation so that joint R & D would not be unlawful.

6.6 Effectiveness and future of government action

It is too early to evaluate the effectiveness of actions by governments designed to improve national competitiveness in biotechnology. Such programmes are only a few years old and it could be premature to come to any firm conclusions at this stage. However, some preliminary observations are presented here.

Studies of the pharmaceutical industry such as that carried out by the OECD [93], have indicated that those countries with a high capacity for innovation in the 1970s were broadly the same as those with such a capacity in the 1950s. In general, one would expect countries with a strong pharmaceutical industry, and strong scientific base to excel in biotechnology as in other related areas. The availability of venture capital and other start-up financing and an entrepreneurial culture are other important factors. All of these factors have contributed to the present pre-eminence of the US in the new biotechnology industry.

However, the future role of government in relation to the biotechnology industry is likely to be influenced in the United States by the wider debate on industrial policy. Proponents of industrial policy such as Robert Reich and Ira Magaziner have argued for an active government role in developing important new industries. They also support full enforcement of United States trade laws, non-tariff restraints, anti-trust policy, and 'targeting' of key industries in the Japanese manner [33,75]. The issue has become highly politicized and at present it seems unlikely that a centralized industrial policy will be implemented in the near future. However, a range of issues are likely to be addressed by United States government agencies without the ideological 'baggage' of being part of an industrial policy. The first of these is co-ordination and reform of the bureaucracy to improve or maintain the competitive position of the US. The shortness of the life of a patent under the regulations is one major issue which may be addressed. Others include the rationalization of regulatory functions between FDA, EPA and USDA and the legislation affecting the export of non-approved drugs. The Cabinet Working Group on Biotechnology recommended in late 1984 the establishment of four new independent expert committees, one each in the FDA, EPA, USDA and National Science Foundation. The committees would be co-ordinated by a Biotechnology Science Board reporting to the Assistant Secretary for Health. In the area of anti-trust legislation it is interesting to note that no action under such legislation has been taken against MCC, the electronics R & D consortium located at Austin, Texas. This may indicate some relaxation in the official position on such R & D consortia.

The situation outside the United States is considerably different. While some countries such as the United Kingdom may attempt to emulate the financial and risk-taking environment within the US, they also engage in co-ordinated development plans designed specifically to strengthen their biotechnology companies. State activity appears to have had a significant role in the period 1980–85 in the emergence of the British biotechnology industry. The central co-ordination provided by the Department of Trade and Industry has rationalized the activities of state agencies, while the provision of state venture capital through the British Technology Group has contributed to the establishment and development of a number of NBFs, most notably Celltech. In France, government programmes in biotechnology appear to be less successful. There are shortages of skilled personnel and French industry has adopted a somewhat conservative attitude despite the government's 'Mobilization Programme'.

In Japan government identification of biotechnology as a major priority has contributed to the major developmental effort currently under way in industry. One of the key policy instruments which seems to be emerging in OECD countries (including perhaps the United States) is the R & D consortium. Such consortia can achieve economies of scale in R & D, more rapid innovation due to pooling of resources and rapid technological diffusion. There are a number of types of consortia. There may be only minimal state involvement, as in the Leicester Biocentre, or major government funding as in the Japanese 'next generation' project under MITI. The research activity may be long-term and basic (as in Leicester) or oriented towards a general improvement in a range of strategically important technologies (MITI Project) or directed towards the common development of process technology.

In summary, government activity designed to enhance national competitiveness will intensify over the next few years. As this activity will provide improved infrastructure and eliminate or reduce regulatory barriers, it will contribute to shaping the environment and competitive activity of the biotechnology industry. National governments will increasingly adopt the role of 'coach', training and supporting their biotechnology teams for the international field.

7 Company case studies

7.1 Genentech

Origin and outline of development

Genentech is generally regarded as the leader of the new biotechnology firms. The company was founded in 1976 to develop commercial applications of recombinant DNA technology by an industrialist Robert Swanson and biologist, Herbert Boyer. Original capital of $100,000 was provided by Kleiner & Perkins, a venture partial partnership in San Francisco. A year later a second round of financing was provided by the Mayfield Fund. Subsequent equity capital was raised in 1978 and in September 1979, Lubrizol purchased 25,000 shares for $10m in cash.

In August 1980 Genentech was one of the first biotechnology companies to go public and there were subsequent refinancings since then. In 1977 the company announced the successful bacterial production of the brain hormone somatostatin. This was an important milestone in the industry as it was the first useful product to be successfully produced by recombinant DNA methods. Genentech followed this up with other successes. In 1978 it cloned human insulin (in a development in association with Eli-Lilly) and in 1979, human growth hormone. In 1980, the year it went public, Genentech produced the hormone thymosin alpha, proinsulin, fibroblast and leukocyte interferon. In 1981 successful technical achievements included the cloning of immune interferon, bovine growth hormone and leukocyte interferon D in genetically engineered yeast. During 1981 the NIH granted approval for large-scale production of five Genentech products including human calcitonin, leukocyte A interferon in yeast, leukocyte D interferon in yeast, porcine growth hormone, and a surface antigen against foot and mouth disease.

During 1982 Genentech's first major recombinant DNA product human insulin was approved for manufacture and marketing by the

FDA by Eli-Lilly under licence from Genentech. The company entered into agreements with Kyowa Hakko and Mitsubishi Chemical Industries to develop and market tissue plasminogen activator in Japan and possibly in other East Asian countries. The company also announced an agreement with Mitsubishi Chemical to develop human serum albumin produced by Genentech scientists. Genentech Clinical Partners Ltd raised $55.6m for clinical testing and developing of alpha interferon and growth hormone. Genentech extended its product development, clinical trial and commercialization effort in 1983 during which immune interferon was in clinical trials. A new R & D limited partnership, Genentech Clinical Partners II, completed a private placement of $34m for clinical testing and development of tissue-type plasminogen activator.

In 1984 Genentech began the first clinical testing of human tissue-type plasminogen activator for dissolving blood clots and two important technical achievements were recorded early in the year including the successful cloning of Factor VIII for use by haemophiliacs and the production of a new natural anti-cancer substance, lymphotoxin.

Genentech's revenue over this period has been from royalties, contract R & D and interest on capital. The company has recorded a high growth in revenue. In 1977 revenue amounted to just $26,000. By 1979 this was $3.4m and in 1980 $9.0m. Revenues for 1981, 1982 and 1983 were $21.3m, $32.6m and $47m. Revenue for 1984 was $69.8m.

Financing

A company such as Genentech, which as of 1984 did not market its own products, is financed in quite a different manner from a conventional company. For example, a typical pharmaceutical company spends about 12 per cent of revenue on R & D. Genentech spends 76 per cent of its revenue on R & D. The company's operating finances are somewhat similar to an R & D department of a large pharmaceutical company with the difference that the R & D department is subsidized by the large product sales of the manufacturing divisions.

In October 1980 Genentech went public with shares offered at $35. Within 20 minutes the share price had jumped to $89 before subsequently settling down. (On 15 July 1983 the price per share was $46.75.) On 31 December, 1981 total shareholders equity was $53.132m.

During 1982 Genentech raised about $100m in cash and commitments bringing the total assets to $101.2m by 31 December 1982. Corning Glass Works bought $20m of Genentech stock in a private placement as did a group of four Swedish organizations, Alfa-Laval AB, an equipment supplier for the biotechnology industry, the Wallenburg Foundation, a private philanthropic foundation, AB Fannyudde, an investment company affiliated with Volvo, and a Swedish Investment Company, D. Carnegie and Co.

Genentech also used the mechanism of the R & D limited partnership to raise $55.6m in 1982 for R & D and clinical trials on human growth hormone and gamma interferon. As described above R & D limited partnerships allow a company to engage in R & D activities without financing them from borrowings or retained earnings. The first Genentech R & D limited partnership, called Genentech Clinical Partners, was completed in December 1982. In 1983 Genentech Clinical Partners II was formed and raised $34m to finance human clinical testing and development of tissue-type plasminogen activator for dissolving blood clots. The company expects to manufacture and market tissue plasminogen activator in the United States under an agreement with the partnership. In late 1984 Genentech formed a third partnership, raising $30.2m.

On the operational side, revenues for the company have grown very rapidly (from $9.0m in 1980 to $47m in 1983). Operating revenues represented about 70 per cent of total revenue in 1981 and increased to about 90 per cent in 1982 and 1983. Unlike some other biotechnology companies, Genentech has recorded a small net income from 1979. In 1983 the net income was $1.128m with total costs of $45.537m. Costs have grown significantly due largely to the increases in the numbers employed (mainly in R & D). Thus at year end 1980 there were 166 full-time employees. This had increased to 543 by year-end 1983.

In 1981 the company purchased about 230 acres of land for $10.3m near its present facilities to help meet future space requirements. In 1982 the company spent $17.6m on capital expenditure and a further $17.1m in 1983. Genentech's 74,000 square feet of bulk manufacturing plant became operational in 1983 and is expected to meet requirements for manufacturing facilities into the late 1980s.

Agreements with other companies

While Genentech plans to manufacture and market its own products in the near future, it, like other NBFs, licensed its early products for manufacture by established pharmaceutical companies. Thus Eli-Lilly has world-wide rights to manufacture and market Genentech's recombinant DNA human insulin. Genentech also has a joint development contract with Hoffmann-La Roche for the production of leukocyte and fibroblast interferons. Hoffmann-La Roche is conducting clinical tests and Genentech will supply part of Hoffmann-La Roche requirements and receive royalty on sales.

Genentech has also entered into a number of joint ventures. In 1982 the company formed a joint venture, Genencor, with Corning Glass Works to manufacture and market industrial enzymes. At present the company markets only enzymes made by traditional technology (pectinase and protease), but it is carrying out R & D into the applications of genetic engineering to enzyme production. In 1983, Genentech formed a joint venture with Hewlett Packard to develop instrumentation for use in biotechnology. The joint venture is investigating both hardware and software needs of the biotechnology industry. Genentech has a minority interest in the new company, HP Genenchem. A third joint venture, Travenol-Genentech Diagnostics was formed with Travenol Laboratories. The venture is for the purpose of developing, manufacturing and marketing human clinical diagnostic products. Genentech has a marketing agreement with KabiVitrum giving the Swedish company world-wide (except United States) marketing rights for Genentech's human growth hormone. The company is developing its animal health product, bovine interferon, in association with Granada which is testing the product. The bovine interferon will be marketed by the company's agricultural market group Genentech Ag. under the trade mark 'Interceptor'.

Strategy

Genentech's strategy has been to concentrate on human and animal health care and to evolve towards becoming an integrated company in this area. The company has so far been quite successful in a number of areas. It has succeeded in raising sufficient funds through public share offerings, private venture placements and R & D limited partnerships to allow it to pursue its aim of becoming the industry leader. The productivity of its research personnel and the excellence of its managerial team have already led to it being regarded as the leading new biotechnology firm. The company consequently enjoys a high profile and this contributes to its continuing ability to raise funding. The company's reputation also assists it in contract R & D and unlike many other biotechnology companies Genentech has been able to finance its operating costs from revenue in each year for the past few years.

Genentech's stated strategy is to become a fully integrated and independent pharmaceutical company. The President, Robert Swanson, has predicted that Genentech will be a '$1 billion company by 1990'. To achieve this Genentech will have to acquire all of the other business functions associated with a pharmaceutical company, since at present it resembles only the research arm of a major pharmaceutical company. The current strategy is to add on the other functions: manufacturing, marketing, and clinical and regulatory expertise. In terms of business development there are three stages in the company's evolution:

1. contract R & D;
2. licensing own products to other companies;
3. in-house manufacture and marketing.

While contract R & D is continuing, the company plans to market its human growth hormone under its own name. This will be followed by three other of four priority products including tissue plasminogen activator for dissolving blood clots, gamma interferon and bovine interferon. The company is initially concentrating on four areas of pharmaceuticals, immunology, endocrinology, cardiovascular agents, oncology and virology. Genentech plans to market these products to medical specialists based at 900 major hospitals around the United States. This can be achieved by using a small sales force which the company can build up over a number of years.

Genentech has taken a number of steps towards realizing these objectives. The proportion of staff represented by R & D personnel has decreased and now accounts for 31 per cent of employees (170); 'process sciences and manufacturing' employ 41 per cent of Genentech's staff (as compared with 24 per cent in 1982). 'Clinical research and regulatory affairs' staff have grown to 217. Research physicians employed by the company recruit clinical investigators to conduct trials at research hospitals and also write clinical study protocols in association with the FDA. The 'regulatory affairs' department is also being expanded. In marketing the company has hired a national sales manager and is initially recruiting a sales force of twenty people.

Genentech possesses a very modern 74,000 square feet manufacturing facility completed in 1983 which is expected to meet its needs into the late 1980s. The facility supplied field-trial quantities of bovine interferon and clinical-trial quantities of tissue plasminogen activator, human gamma interferon and human growth hormone in 1983.

The financial strategy has been to operate at or near breakeven point by financing current costs from operating revenue. The major financial problem faced by the company in its bid to become an integrated pharmaceutical company was (and is) the cost of financing clinical trials. To date $120m has been raised through three R & D limited partnerships. However, much additional funding will be required over the period from now to 1990 to bring all of the products currently envisaged to the market.

Genentech has consistently placed emphasis on personnel and has a strong managerial team. The President, Robert Swanson, has a high profile within the industry. The company has recruited many top scientists and places emphasis on the creative environment within the company.

Genentech's strategic focus has been on human and animal drugs but ongoing research activities within the company have led to other innovations with commercial potential. To exploit those fully while not distracting effort away from the central mission, Genentech has formed joint ventures with other companies. Thus, industrial enzymes, biotechnology equipment and diagnostics have been 'spun-off' into joint ventures with minority interests for Genentech. Genentech's strategy can be summarized as follows:

- focus on human/animal health care;
- development of all functions of an integrated pharmaceutical business;
- marketing of own products to hospital specialists in four priority areas, endocrinology, immunology, cardiovascular agents and virology;
- operating expenses financed through revenue from contract R & D and royalties;
- clinical trials financed through R & D limited partnerships;
- non-drug research 'spin-off' into joint ventures.

The strength and weaknesses of Genentech's strategic position are evaluated in Chapter 7.

Sources: [18, 42, 46, 50, 55, 58, 59, 65–67, 97, 98].

7.2 Genex

Origin and development

Genex was founded in July 1977 by Robert F. Johnston, a venture capitalist, and J. Leslie Glick, a biochemist who was previously with Associated Biomedic Systems. Initial financing was provided in 1978 and 1979 by Innoven Capital Corporation, a venture capital organization and additional financing was provided from the Koppers Company in 1979 and 1980. During 1982, the company went public and total stockholders' equity was $26.8m at the end of 1982.

Genex has concentrated on the area of specialty chemicals and enzyme technology in contrast to many other biotechnology companies which have emphasized health care applications. A major part of Genex's activity in the early years was in developing contract R & D and technology assessment services. Genex carried out contract R & D in pharmaceuticals for a number of other companies including projects relating to interferons, interleukin-2, tissue plasminogen activators and blood factors.

The company's technology assessment services advise other companies on the feasibility and production economics of using genetic engineering to make their products. The first revenues from this activity were obtained in 1978. In 1979, the company grew to nineteen employees and Genex established laboratories in Rockville, Maryland. In 1980 Genex began a long-term research project for Koppers in aromatic chemicals and Koppers made additional investments in the company. By the end of 1981 Genex had grown to 191 employees and the company purchased and renovated a facility in Gaithersburg, Maryland for expanded research space and pilot plant.

In 1982 Genex had its first public stock offering. The company was also successful in cloning the gene for rennet (used in cheese making) and also sold its first product, aspartic acid.

In 1983 the company accelerated its transformation from an R & D services company to a fine chemical manufacturer. The company supplied Searle with phenylalanine made under a toll manufacturing arrangement with Cell Products Ltd. Genex purchased a 280,000 square feet production facility at Paducah, Kentucky for the production of phenylalanine and aspartic acid and other products. Substantial product sales ($1.8m) were recorded during 1983, increasing to $20.6m in 1984.

In 1984 Genex transferred production of L-phenylalanine to its Paducah facility. Product sales for the first six months represented 69 per cent of the company's total revenue compared with less than 1 per cent of the company's total revenue in 1983. The company introduced an enzymatic cleaning product, 'Proto', for dissolving hair in drains.

Financing

The Koppers Company played an important initial role in financing Genex and invested in the company in 1979 and 1980. The company's initial stock offering in September 1982 raised the total stockholders' equity from $15.2m (end of 1981) to $26.8m (end of 1982). Working capital in 1982 increased by $11.1m primarily as a result of funds raised on the share offering ($16.9m in 1982). In April 1983, Genex completed its second public stock offering and raised an additional $16.3m. At the end of 1983, total stockholders' equity had increased to $37.8m. Genex has not entered into any R & D limited partnerships to date.

On the operational side, Genex has used its working capital obtained from the share offerings to finance large losses in 1982 and 1983. In 1982, the company recorded revenues of $6.1m and expenses of $11.7m i.e., a net loss of $5.6m. In 1983, Genex recorded a loss of $5.4m. These losses were caused by an accelerated programme of proprietary R & D, process scale-up and construction of pilot plant. The cell products facility for manufacturing phenylalanine also had a $1.5m start-up expense. Given this high level of losses, the company financed its investment in a major production facility at Paducah, largely through debt.

Despite the temporary losses in 1982–83, the company's finances looked reasonably sound in the first three quarters of 1984. Product sales (largely phenylalanine) rose from $9,400 in 1982 to $1.76m in 1983 and $9.8m in the first half of 1984. However product sales in the fourth quarter were only $100,000 leading to a net loss in the quarter of $7.5m. Currently 75 per cent of Genex's total operating revenue is derived from product sales, a situation unmatched in any of the other new biotechnology firms. The company recorded a net loss of $7.4m in 1984 and in December 1984 made 54 employees redundant. Genex appears to be experiencing price pressure from its sole phenylalanine customer G.D Searle and cost over runs at its manufacturing facility.

Agreements with other companies

Genex has (or had) a variety of arrangements with other companies. These include contract R & D arrangements, consultancy services, joint research, and a manufacturing agreement.

In the area of contract R & D, Genex completed contracts with Green Cross & KabiVitrum in 1983 for the development of a micro-organism that produces human serum albumin and also completed a contract with Schering for an amino acid-producing organism. Genex is currently involved in a contract with Yoshitomi Pharmaceutical Industries of Japan to produce interleukin-2 through genetic engineering techniques.

As well as carrying out contract R & D, Genex also offers consultancy in which it undertakes feasibility studies of potential biotechnology applications including advising on research strategies and production costs. It also offers services in the area of bioprocess design and development.

Genex entered into a five-year, $16.5m R & D contract with Bendix, on 1 January 1983 to develop protein engineering technology. The objective was to develop the basic and applied technology necessary for constructing novel proteins with desirable characteristics. This technology could be coupled with existing genetic techniques to lead to the creation of advanced products of known physical, chemical and biological specificity. The intellectual property rights for this joint programme were assigned to a partnership termed Proteus Associates with Bendix owning 70 per cent and Genex 30 per cent. Under the agreement both companies would be granted royalty-free licenses from Proteus. In the early part of 1984, after the acquisition of Bendix by Allied, it became apparent that the Bendix/Allied partner wished to pull out of Proteus and in July 1984 Genex and Allied signed an agreement that transferred complete ownership and rights of the partnership to Genex. Genex is now continuing with the protein engineering programme. The other major intercompany agreement was with Cell Products whereby that company manufactured phenylalanine for Genex. This process has now been transferred to Genex's own Paducah facility.

Strategy

As Genentech plans to become a pharmaceutical company, Genex aims to be a fine chemical company. The company targeted the area of fine chemicals made through recombinant DNA and other techniques at an early stage unlike many of the other NBFs which concentrated on health care. Genex has done some R & D for clients on therapeutics, but this has no relation to its central business. Genex's strategy has been to concentrate on applications of the new technology which are less costly and involve much less regulation and can therefore be brought to market much faster than pharmaceuticals. The company has identified a number of important specialty chemical markets where it hopes to compete. These include amino acids, enzymes, vitamins and other chemicals.

Its amino acids are currently the most advanced products, with phenylalanine selling well. The company also plans to be a major supplier of aspartic acid (the other ingredient of aspartame) and both of these amino acids will shortly be manufactured in-house. Genex is also developing three other amino acids, tryptophan, threonine and serine. The first two would be used as animal feed additives and serine would be used as a raw material for the production of tryptophan. It is likely that each of Genex's amino acid products will have a different marketing strategy e.g. tryptophan for mineral premixes can be sold to milling companies rather than directly to farmers.

Genex lays special emphasis on enzymes and in 1984 launched an enzyme-based drain cleaner, 'Proto'. The product is aimed initially at the $200–300m American institutional cleaning market and eventually at the domestic market which is worth $150m. The product dissolves hair in drains and avoids use of caustic agents such as sodium hypochlorite. Genex is developing a range of enzymes for various applications in food processing. The company has cloned the gene for calf rennet used in the coagulation of milk and is currently involved in process scale-up and the testing of the enzyme in cheese manufacture. Genex believes that recombinant rennet will offer advantages over conventional rennet obtained from calf stomachs in both price and quality. Genex also plans to enter the market for starch enzymes with alpha amylase and glucoamylase. Another market which has been targeted is the detergent enzyme market, for which the company is currently developing a process for alkaline protease. The American market for detergent enzymes is considerably underdeveloped in comparison with the European market and Genex hopes to capitalize on this.

The company is also developing new processes for vitamins. Vitamin B12 currently has a high production cost and is very expensive ($2,000–3,000 per pound). Genex is attempting to reduce this production cost through the use of genetic engineering. The company is also interested in the production of vitamins A, C and E but these are likely to be more long-term goals.

In choosing to enter these markets Genex will be competing with established companies such as the Japanese company Ajinomoto in amino acids and Novo and Gist Brocades in enzymes. Genex's marketing strategy will combine a number of different approaches for different product areas. Some enzyme products can be distributed in the United States by a small marketing force. In the waste treatment enzyme sector, Genex will use marketing links with one or more specialty chemical firms. For the drain cleaner product, Genex plans to form arrangements with distributors in the institutional janitorial market. For distribution in Europe, the company will have to rely on European distributors in various markets. In the area of detergent and starch enzymes the marketing will be relatively simple since there are only three large soap powder companies and about six corn wet millers in the United States.

The company's financing to 1984 has reflected the movement into full-scale manufacturing. This has involved large expenses in scale-up and purchase of pilot plant and full-scale manufacturing facility. Unlike many other NBFs Genex has used some debt financing and at a time when the company was making heavy losses. This was to fund the full-scale production facility and it would appear that the heavy investment in scale-up in 1981–83 and the costly investment in manufacturing facilities were justified by the early market entry of Genex's products. The company has pursued a high risk financial strategy over the last two years and has now achieved an important strategic lead over many competitors in the area of full-scale production.

Genex has a strong management team with extensive experience in the pharmaceutical and chemical industries. The Chairman and Chief Executive Officer, J. Leslie Glick has a high profile within the industry.

One notable feature of Genex's strategy, has been its planning for its own technological obsolescence. Genetically engineered products such as enzymes represent the 'first generation' product. Production of the 'second generation' enzymes will involve a range of technologies including genetic techniques and also protein engineering. Genex's investment in the protein engineering area is designed to give the company the ability to compete in these second generation products. Similarly the company is investing in research into immobilized enzyme bioreactors and is attempting to develop proprietary enzyme bioreactors which will be protected by patent.

Genex's competitive position in the near future will be based on the superior production economics in the manufacture of established specialty chemicals obtained by a proprietary, high-level expression and secretion system in the bacterium *B.subtilis*. In the longer term Genex's competitive strategy will rely more on performance, i.e., the development of specialty chemicals with improved functional characteristics and on 'new' specialty chemicals which may be analogues of existing chemicals or entirely new products serving a similar function. The company has an excellent market focus and is well positioned to compete effectively in a number of specialty chemical markets. Nevertheless, one serious question mark over the immediate future is its contract to supply phenylalanine to Searle. The latter company plans to make phenylalanine itself and should Searle withdraw its contract suddenly, the impact on cash flow in Genex would be very significant.

Sources: [6, 38, 42, 63, 64, 80, 97, 98, 111].

7.3 Cetus

Origin and development

Cetus was founded in 1971 by Peter Farley and Ronald Cape. Unlike Genentech or Genex which have concentrated on applying biotechnology to a particular sector, Cetus developed a wide range of interests in diverse industries including energy and biomass, sweeteners, fine chemicals, bulk chemicals, agriculture and pharmaceuticals. From the beginning the company has sought a variety of industrial partners with which to commercialize its R & D programmes. These have included companies such as Amoco, Shell, Socal and Schering.

The company's revenues grew from $2.4m in 1977 to $9.8m in 1981, the year in which Cetus went public. Cetus' annual report from 1981 lists a wide range of R & D programmes including:

- new enzymes;
- conversion of ethylene and propylene to their oxides and glycols;

- development of a process for high purity fructose;
- plant genetics;
- fibroblast interferon;
- hormones;
- diagnostics;
- monoclonals for rheumatism/arthritis;
- vitamins;
- antibiotics.

Cetus began a collaboration with Standard Oil of California (Socal) in 1979 to develop a process for manufacturing high-purity fructose from low-cost, corn-derived glucose. In May/June 1982 Socal decided to pull out of this programme. This led to the abandonment of the programme by Cetus with the lay off of forty staff and a major reorientation of its research effort. The company subsequently concentrated on three main areas, health care (especially cancer therapeutics and diagnostics), agricultural products, and industrial processes and products. It is not possible to present even an outline of the development of the company without discussing its relationship with other companies and so this will be presented here before a discussion of the financing.

Agreements with other companies

Cetus' agreements include contract R & D and licensing, joint ventures and marketing agreements. In common with other NBFs, Cetus has carried out a range of contract R & D and has developed a number of products in association with other companies. Cetus developed improved strains of bacteria for making the antibiotics Sisomycin and Netromycin which are being marketed by Schering-Plough. The company developed a monoclonal antibody for use in low back pain diagnosis which is being marketed by Cappel Laboratories. Other achievements include the marketing by Norden Labs of a vaccine for the prevention of scours in newborn pigs. These products are currently earning royalties for Cetus. The company signed a research agreement with Shell Oil Company in June 1980 to develop microbial methods for interferons. The company's product 'Betaseron' (recombinant beta interferon) had completed Phase I clinical trials in 1984 and Phase II was about to begin. Appplications of the product to treatment of hepatitis and influenza are also being investigated.

Cetus has formed two subsidiaries to exploit various applications of its technology. Cetus Immune Corporation was founded in December 1980 to concentrate on applications in immunology. Its laboratory facilities were opened in Palo Alto in May 1982. Cetus Madison Corporations was founded to concentrate on agricultural applications especially the genetic engineering of plants and is based in Middleton, Wisconsin. In June 1984, Cetus announced a joint venture 'Agracetus', with W.R. Grace. Agracetus will develop, manufacture and market agricultural and veterinary products. Grace owns 51 per cent of the joint venture and Cetus is contributing the assets of its agricultural subsidiary, Cetus Madison. Grace is expected to contribute $60m to Agracetus. Cetus also formed a joint venture with Nabisco Brands Inc. in 1984 to apply biotechnology to food and food ingredients. The venture is called Nabisco/Cetus Food Biotechnology Research Partnership. To market its products worldwide Cetus is making a variety of marketing agreements with overseas firms in Europe and Japan.

Financing

Cetus was initially financed by a $2m venture capital investment. Subsequent financing by Socal invested $5m for 10.4 per cent of the company. Cetus had its first public stock offering in March 1981 and the net proceeds were $107.2m — the largest amount raised by any of the new biotechnology firms. In 1983 an R & D limited partnership, Cetus Healthcare Limited Partnership was set up and raised $75m. At the end of 1984 total stockholders' equity was $134.2m. Revenues increased from $2.09m in 1978 to $15.5m in 1981 the year the company went public. Subsequently, revenues more than doubled in 1982 to $32.9m and went to $28.8m in 1983. In 1984 an additional $19m R & D revenue was obtained from the R & D limited partnership, leading to a 60 per cent increase in 1984 revenue over 1983 levels.

Cetus made a net profit of $4.5m and 0.9m in 1982 and 1984 respectively and a loss of $4.6m in 1983. Product sales in 1984 were $2.9m. Investment in property, plant and equipment in 1982 was $22.9m. This was for partial construction of a production plant, expansion of the main R & D laboratories at Emeryville and construction of facilities for two subsidiaries. The financial policy is to operate at, or near, break-even level and to use the proceeds from the stock offering and the funding provided by joint venture partners to finance expansion of programmes until Cetus' own products begin to generate substantial revenue.

Strategy

Cetus is an interesting case study because it has undergone a change in its market focus over the course of its development. The early strategy was to demonstrate technical and commercial feasibility on products and processes before joining with another company or pursuing commercialization itself. The original focus of the company was extremely broad, encompassing a range of industries and sectors. In particular, Cetus hoped that it could rapidly develop new technology in sweeteners and bulk chemicals which would allow rapid growth into a major industry leader. The Chairman, Ronald Cape, expressed the aim of becoming the 'IBM of biotechnology'.

On the fructose project the 1981 Annual Report states

> we anticipate that near-term revenues will be dwarfed in scale by revenues from the markets we plan to enter later in the decade. The most rapidly advancing of these world scale programs is the Cetus process for making high-purity fructose.

When Socal pulled out of the fructose project Cetus found itself unable to support the development costs of such a major project and was forced to abandon it. This also led to a re-organization of the research effort into a smaller number of areas. The 1982 Annual Report emphasizes diagnostics, cancer therapeutics and agriculture and states:

> During 1982 Cetus redirected its strategy of pursuing many programs across the full spectrum of biotechnology opportunities, to one of focussing on a smaller number of programs with a greater degree of effort applied to each one.

The 1984 Report emphasizes three main areas, health care products (especially cancer related), agricultural products, and industrial 'processes and products'. Cetus is currently placing major emphasis on diagnostics and therapeutics for cancer and sees itself as a major anticancer company of the 1990s. Cetus is developing tests for a variety of types of cancer using monoclonal antibody and DNA probe technology. The company will introduce two tests for prostate cancer in 1985 and another test is being developed for gastro-intestinal cancer. Cetus is also conducting research into oncogenes.

In therapeutics the company is pursuing research into interferons, lymphokines and immunotoxins. The company's interferon product 'Betaseron' is undergoing clinical trials and Cetus hopes to submit an application to the FDA to market 'Betaseron' in 1987.

The lymphokine interleukin-2 (which promotes growth of cells involved in the immune response) has been cloned and expressed by company scientists and early results suggest it may have a strong anticancer effect. The Cetus interleukin-2 is a modified version of the natural molecule which the company hopes will provide it with a proprietary patent position. Cetus will apply to the FDA for approval to market interleukin-2 in 1988.

By linking a strong cell toxin to a monoclonal targeted against a cancer cell it is possible selectively to deliver the poison to the cancer cell and kill it. Such a combination, termed an immunotoxin is currently being developed at Cetus and could be marketed in 1989.

Agracetus is investigating applications of recombinant DNA and plant techniques such as regeneration of entire plants from tissue culture to engineer desirable traits in plants. The company has identified genetically engineered plants and crop treatments as one of its priorities and hopes that this technology will have a major impact on the fertilizer, herbicide and insecticide markets.

In the past Cetus had not required a significant marketing and regulatory expertise since these functions were carried out by the industrial partners with which it was developing products. In line with its emphasis on health care, the company has built up a strong group for the handling of pre-clinical and clinical trials and for dealing with regulatory affairs. In marketing, the company will pursue a wide range of arrangements. The joint ventures formed in the areas of plant biotechnology (Agracetus) and food biotechnology (with Nabisco) will market Cetus' products in these areas. Some products are being marketed directly by Cetus itself including the diagnostic test for cytomegalovirus and an automated liquid sample handling device 'Pro/Pette'. The last product has been a major contribution to the product sales in 1984 with $2.4m of the equipment sold. World-wide distribution of products will be achieved through cross licensing and partnership arrangements with other companies (Cetus planned at one time to open a British subsidiary). Cetus has maintained an innovative climate within the company. There has traditionally been a high emphasis on scientific independence and this has led to much creativity and a stimulating environment for Cetus scientists. However, this scientific freedom may have gone so far that the firm has lacked commercial focus and failed to develop a coherent business plan soon enough.

Cetus has a number of strengths at present. It has a strong anti-cancer programme which is likely to come up with some products, even if this takes some years. Meanwhile, it has adequate funds available for lengthy and costly product development and commercialization. The diagnostic tests it is developing should also provide additional short-term cash flow until its major therapeutic products are introduced.

In summary Cetus provides an interesting case study of strategic re-orientation within an NBF from a broad-based company to a more focused one. The proximal cause of this was the failure of the chemical venture with Socal. However, the company seems to have used this as the basis for a complete rethink on its strategy. The strategy today is not dissimilar to that of Genentech, Genex or Biogen, although Cetus is behind Genentech in terms of forward integration into health care markets. However, for most of its life, Cetus has exemplified the broad-based NBF attempting to enter a number of sectors simultaneously. The example of Cetus is used here, therefore, to illustrate this broad-based strategy although the company has since moved on from this original policy. *Sources*: [7–9, 14, 27, 38, 42, 47, 97, 98].

7.4 Centocor

Origin and development

Centocor was set up in 1979 by Michael Wall and Hubert Schoemaker on the campus of the University of Pennsylvania at Philadelphia. Through 1980 and 1981 the company concentrated on arranging finance and establishing contracts with academic researchers. Centocor concentrates on human health care, in particular cancer diagnostics and therapeutics. In 1981 the company joined with FMC Corporation to form a joint venture for developing human antibodies. The scope of this joint venture was subsequently expanded in 1982 into the area of immunoregulation. In 1982

Centocor had its first public stock offering and raised $21m. In 1984 the company raised $15 in a R & D limited partnership.

The company currently has eight diagnostic products at various stages of development and marketing as well as a range of others at earlier research stages. In 1983 Centocor introduced three diagnostic products: an assay for hepatitis B based on monoclonal antibodies was distributed by Warner-Lambert Corporation in Europe and two Mab-based cancer tests, one for gastrointestinal cancer (CA-19-9), and one for ovarian cancer (CA-125), were also introduced to the European market and for experimental use in the United States while awaiting FDA approval.

In 1984 Centocor introduced an assay for gamma interferon which is being used for research purposes to test the immune status of immuno-compromised patients. A second gastrointestinal test for early detection of lower tract tumours has also been developed. Also in the pipeline are tests for liver and breast cancer which are expected to reach the American market in 1985/86. Centocor has also obtained licences to two oncogene patents from Applied Biotechnology Inc. and the Massachusetts Institute of Technology and expects to market a diagnostic kit based on Mabs against oncogene proteins in late 1985 in Europe and the Far East. The company is also concentrating on longer term research in the area of *in vivo* diagnostics and cancer therapeutics using monoclonal antibodies. It is conducting research into *in vivo* imaging of gastrointestinal and ovarian cancers in collaboration with university centres. Diagnostic imaging is also being applied to detect the extent of heart damage following myocardial infarction by use of an anti-cardiac myosin antibody developed by the company.

Financing

Total shareholders' equity increased from $4.6m in December 1981 to $22.8m in December 1982 as a result of the public stock offering. By then the company had an accumulated deficit of $5.5m. Revenue (mostly from contract R & D) has increased from $153,000 in 1980 to $7.4m in 1983. Product sales of $1.2m were recorded in 1983. The majority of the company's R & D revenues have been obtained from reimbursement to Centocor for research conducted for Immunorex, its joint venture with FMC Corporation. Thus of $2.4m of contract R & D revenue in 1982, $1.7m came from this source. FMC entered into an agreement to fund the first $4.9m of Immunorex's human antibody programme and $7.5m for the initial costs of the immuno-regulation programme. Subsequent contributions to the first programme are being made on an equal basis by FMC and Centocor.

The company has also received payments from the distributors of its products in advance of product sales. Centocor has funded external academic research on its behalf in a variety of institutions. The arrangements with these research institutions involve giving Centocor licences and/or options to license technology resulting from the research and the obligation of paying 4.8 per cent royalties on product sales.

Agreements with other companies and research groups

Centocor has established a variety of agreements with other companies in each of the areas in which it is concentrating. These agreements include joint development and marketing. In therapeutics for cancer, Centocor works with FMC through its joint venture Immunorex Associates. In 1983, the company entered into a joint development programme with Hoffmann-La Roche for the development of therapeutic products using non-human antibodies. In the area of diagnostic imaging, the company works with a number of partners in order to use their experience and resources. These include Medi-Physics, a Hoffmann affiliate in the United States, Commissariat à l'Énergie Atomique in Europe and Nihon Medi-Physics in Japan, a joint venture of Nippon Roche K.K., Sumitomo Chemical Company and Sumitomo Corporation. The company is offering its cancer diagnostic kits directly for research use and in April 1984 signed an agreement with the leading world diagnostic company, Abbott for marketing these two cancer diagnostic kits. Centocors' hepatitis B test which is the first to incorporate monoclonal antibodies is being sold by Warner-Lambert and other distributors.

In addition to agreements with companies Centocor has a number of research agreements with leading biotechnology and clinical institutions including, the Wistar Institute, the Dana Farber Cancer Institute, Memorial Sloan Kettering Institute, and Massachusetts General Hospital. The company uses its contact with medical and research centres to conduct research and testing on its behalf and licenses antibodies which can subsequently be developed into assays, proprietary to Centocor.

Strategy

Centocor has focused on a well-defined area covering cancer diagnostics, therapeutics and other diagnostics. As a small company the strategy has been to obtain the maximum leverage through accessing a range of business functions outside the company. Thus Centocor is relying not only on other established companies to market its products, but also on external academic researchers to amplify the R & D resources available to it. These relationships with outside institutions form an important part of the business strategy and the company therefore will pay royalties as well as receiving them from other companies.

The key element of the strategy is to focus on products which can be brought to market fast (i.e. diagnostics) and generate cash-flow while at the same time pursuing long-term and costly development in partnership with larger firms. The company sees itself supplying 'the key components of a new health care system' to other companies in a manner analogous to Intel's supplying chips to electronic system manufacturers. Consequently, it has been able to demand a high royalty on technology which it licenses (20 per cent). The company does not intend to become a full-scale pharmaceutical company in its own right but will work through the licensing and joint venture mechanism with the established pharmaceutical firms.

Technically the company has strong resources and employs over 100 people, excluding the external technical resources. The areas chosen for product development are likely to be very high growth such as monoclonal therapeutics for cancer and *in vivo* diagnostic imaging and the use of the resources of large corporations for product development and marketing will greatly facilitate the company's ability to compete. The policy of commercialization of the research results performed outside the company speeds up the process of new product introductions. The company has agreements with more than fifteen different research groups world-wide.

The issuing of non-exclusive licences to marketing partners in order to cover the world market is an important element of its strategy and it has seen that the competition in the markets which it has selected will be global. Other NBFs such as Hybritech, Xoma, Immunomedics and Cetus as well as major established companies will compete intensely in cancer diagnostics. At present Centocor appears to be technically ahead of most other companies in the field and its innovative strategy will undoubtedly make it a significant force in this area in the near future.

Sources: [10, 25, 29, 48, 98].

7.5 Celltech

Origin and Development

Celltech, the British company, was established in 1980 as a joint venture between the state and private enterprise. Gerard Fairtlough of the British National Enterprise Board played a key role in setting up the company and subsequently became Chief Executive. The National Enterprise Board took 44 per cent of the equity and four private investors, British and Commonwealth Shipping, Prudential and Midland Bank and the Technology Development Capital part of the Finance for Industry Group, controlled the rest.

The firm was set up to act as the British challenge to American industry leaders such as Genentech. The United Kingdom had already missed out in the 1970s when the invention of monoclonal antibodies was made at Cambridge and no patent was taken out. This allowed a world-wide diffusion of the technology and led to the creation of many NBFs in the United States.

Celltech originally concentrated on the application of two new technologies, recombinant DNA and monoclonals, to health care although some recombinant DNA work is in other areas such as calf rennet. The emphasis, however, was on the monoclonal antibody area. The company introduced its first products within one year. These were a reagent for the assay of alpha-interferon and a chromatographic medium for obtaining commercial quantities of high purity interferon based on monoclonal antibody technology. By the end of 1982, Celltech had launched two other products, monoclonal antibodies for blood group typing. In 1983 the company entered into a joint venture with Boots to form Boots-Celltech Ltd. which is developing and marketing diagnostic kits based on monoclonal antibodies.

During 1983, the Technology Development Capital and National Enterprise Board equity holders decided to realize part of their investment in the company and Biotechnology Investments the biotechnology investment trust set up by N.& M. Rothschild took an 11 per cent share in the company. Celltech originally had a close relationship with the United Kingdom Medical Research Council and had first option on the commercialization of research carried out in MRC laboratories. At the end of 1983, this relationship was altered so that Celltech had exclusive rights only on directly funded work and first option on areas where Celltech had been or would soon be active in commercializing Medical Research Council work.

Financing

Celltech was originally established with £12m supplied by the initial investors. During the reorganization of the equity holders in 1983, Celltech issued additional shares and the net proceeds of share issues and calls aggregated £10.6m raising the cash balance of the company from £4.3m to £11.6m. The total operating revenue increased from £71,000 in 1981, the first year of operation, to £384,000 in 1982 and £909,000 in 1983. However, the company had an accumulated deficit of £4.8m at the end of 1983 and its operating deficit for 1983 alone was over £3m.

In 1983 Boots-Celltech was set up which involved the sale of the company's diagnostic division. This raised an exceptional profit of £669,000 and depending on the sales performance of the joint venture in the years ahead, an additional £500,000 is expected. In 1984, Celltech finances showed the beginning of a turnaround. Losses at £1.9m were slightly less than the £2m recorded for 1983 but turnover increased 125% to £1.9m. Celltech is continuing to invest heavily in new plant and is commissioning a new factory for the Culture Products Division at a cost of £2.4m, which is due to open in 1985.

Agreements with other companies

These cover contract R & D and licensing agreements, marketing arrangements and joint ventures. Celltech's 'culture products division' carries out bulk production of monoclonal antibodies for other companies. The company possesses two 100-litre fermenters and a 1,000-litre fermenter and can produce 5 kg of monoclonal antibodies annually. This probably represents the world's largest industrial production of monoclonals using *in vitro* methods. Celltech's main competitor among the new companies is the American company Damon Biotech, which uses a proprietary encapsulation method for producing bulk monoclonals. The high cell density involved in this latter process means that a lower bioreactor volume can be used than in the Celltech process which grows hybridomas in suspension in air-uplift fermenters.

In early 1984 Celltech signed a contract with Serono Labs. Inc. for the development of human growth hormone, a drug which has already been developed by Genentech. The Celltech product will be manufactured using mammalian cells.

In 1983 Celltech entered into an agreement with Sankyo Company of Japan for the development of calcitonin and tissue plasminogen activator. The products are expected to go into clinical trials in 1985–86. In 1984 the company entered into two further agreements with Sankyo for the development of macrophage activating factor and tumour necrosis factor.

Celltech also has marketing and distribution agreements with a number of overseas companies including Sumitomo which distributes Celltech products in Japan.

Celltech has two joint ventures, Boots-Celltech Diagnostics Ltd. and Apcel Ltd. The former was started in August 1983 as a joint venture with Boots to commercialize and distribute Celltech's diagnostic products. In its first year of operation, Boots-Celltech Diagnostics Ltd introduced four new diagnostic tests for:

- alpha foetoprotein for foetal abnormality;
- thyroid-stimulating hormone;
- chlamydia;
- respiratory synctial virus.

The joint venture is also preparing to introduce new products in the area of human and bovine fertility. Celltech's second joint venture, Apcel Ltd was formed in 1984 with Air Products Ltd. The new company will carry out contract R & D for the two participating companies and others in the area of industrial microbiology.

Strategy

Celltech has concentrated on the applications of monoclonal antibody technology as have a number of other NBFs such as Hybritech, Centocor, Monoclonal Antibodies Inc. Its strategic focus is in the area of immunodiagnostics and it has rapidly introduced a number of innovative products. As in the case of Centocor, this generates a rapid cash flow which assists in the development of further products. Unlike the above NBFs, however, Celltech also has interests in the application of recombinant DNA technology to industrial microbiology. The joint venture with Boots assists in commercialization and the company plans rapidly to introduce a range of new Mab-based diagnostics and to play a leading role in the diagnostic market. The development of a strong technical capability in large-scale cell culture gives Celltech the following advantages:

- an industrial scale production system for Mabs for its own diagnostic products;
- a production technology for pharmaceuticals developed in partnership with other companies;
- a contract R & D facility for generating additional revenue.

As in the case of Centocor, Celltech's relationship with the external scientific community played an important part in its strategy. Although its relationship with the Medical Research Council has changed, Celltech still enjoys a close collaboration with its laboratories and has first option rights on Medical Research Council research results as described above. The company also has links with academic institutions. Celltech is well positioned to compete in the applications of monoclonal antibody technology to health care. The company is an outstanding example of a European start-up rapidly achieving the technical and commercial success which is associated with American new biotechnology firms.
Sources: [28, 39, 40, 42, 51].

7.6 Eli-Lilly

Origin and development

Eli-Lilly is a major American pharmaceutical company with total sales of $2.7 billion in 1982 (earnings of $412m). In 1984, Lilly's R & D budget was approximately $294m. The company makes pharmaceuticals, medical instruments, agricultural products and cosmetics. In pharmaceuticals, major product groups include injectible antibiotics, oral antibiotics, insulin, cardiovascular drugs and cancer drugs. Lilly has a very large share of the American insulin market (approximately 80 per cent) but does not dominate the European market which is served by its main rival in insulin, Novo of Denmark. In medical instruments, Lilly makes cardiac pacemakers, insulin infusion pumps, cardiac defibrillators, vital signs monitors, intravenous delivery systems and other equipment. In the agricultural products area, the company produces animal antibiotics and herbicides and products to improve weight gain in animals.

Lilly carries out R & D in a wide variety of areas related to its main divisions and their products. In pharmaceuticals the company is investigating new types of antibiotics, improved drug delivery systems, products for the management of blood pressure, allergy, cancer and arthritis, and neuroactive peptides.

The company's first major involvement in biotechnology was in the application of recombinant DNA technology to produce human insulin (until recently only porcine insulin was available for diabetics). Lilly licensed this technology from Genentech which cloned the genes coding for human insulin. Full-scale manufacturing plants were constructed by Lilly and the product 'Humulin' was introduced into the British market in 1983, and subsequently into the American market. Initial sales, over the period from 1983 to end of 1984 were slow (only 1 per cent of British market after seven months) but the American market for Humulin is expected to increase to $30m in 1985. Lilly has faced strong competition in its home market from Novo which formed a partnership Squibb-Novo to market its insulin in the United States. Novo markets highly purified porcine insulin and also its own semi-synthetic human insulin which is obtained by chemical modification of porcine insulin, rather than by cloning the genes coding for human insulin. Novo has teamed up with the biotechnology company, Biogen, to develop its own recombinant human insulin and in late 1984 Novo applied to the Dutch authorities for permission to construct large-scale facilities for production of recombinant DNA insulin. The Novo process involves cloning of the precursor of insulin, proinsulin, and its enzymatic cleavage to produce insulin. This differs from Humulin which involves separate cloning of the A and B chains of insulin and then joining them to give the insulin molecule. (In the human body proinsulin is converted to insulin by cleavage of a peptide connecting A and B chains.) Lilly has also developed its own process for recombinant DNA production of proinsulin.

The development of human insulin illustrates a number of interesting features of the emerging biotechnology industry:

- Human insulin was the first drug developed using recombinant DNA technology. This pioneered the application of recombinant DNA technology to the production of natural therapeutic substances. As such, it represented an important breakthrough for the biotechnology industry. It demonstrated conclusively that the new technology could be successfully scaled up and commercialized and acted as a powerful confidence builder in the industry.

- 'Technology-push' played a major role in its development. Innovations may be regarded as being predominantly influenced by 'technology-push' or 'market-pull', although both factors are almost always present. Recombinant DNA human insulin resulted from the partnership of Genentech's scientific ability with Eli-Lilly's desire to introduce a major innovation in its insulin range. Lilly has claimed in promotional literature that human insulin was developed in response to a forthcoming shortage of porcine glands and a demand for the human product which did not have the immunological properties of porcine insulin. A shortage of porcine glands may indeed be a long-term problem, but this does not adequately explain the major investment in this product at a time when no shortage was evident in the immediate future. Although the recombinant DNA human insulin is superior immunologically to porcine insulin, clinical evidence at present does not appear to support a contention that it is a dramatic improvement in general diabetic therapy. Recombinant DNA insulin does, however, represent a dramatic change in process technology, which suggests a high degree of 'technology-push' in this pioneering innovation, a situation also seen in some other biotechnology products such as Mab diagnostics.
- The competition between Eli-Lilly and Novo illustrates the phenomenon of 'leap-frogging' in technology. Novo did not attempt to emulate Lilly's technology when it decided to produce its own recombinant DNA human insulin. Instead it attempted to 'leapfrog' ahead to a more advanced and cost effective recombinant process. This phenomenon is a common strategy in an area where technical change is rapid and is particularly important in biotechnology as we shall see below (Chapter 8).

Apart from its insulin project, Lilly has invested in the application of recombinant DNA and immunology to the wide range of therapeutic classes mentioned above. In 1984, the company opened its new Biomedical Research Centre at its headquarters in Indianapolis. The company is carrying out research into the application of biotechnology to respiratory and infectious diseases, cancer and arthritis and also into immunology.

Respiratory diseases/infectious diseases

The company will investigate diseases such as asthma, emphysema, pulmonary fibrosis, and acute respiratory distress syndrome. Particular attention will be paid to the role of leukotrienes and immunological factors in lung disease. In the area of infectious diseases, Lilly has targeted respiratory infections, peritonitis, influenza and the common cold.

Cancer research

Lilly's cancer research will explore a variety of possible therapeutic avenues. There will be retroviral studies, and the biochemistry and cell biology of oncogenes and growth factors will be investigated. The company is also looking for new substances which can prevent normal cells becoming cancerous, and others which might inhibit tumour blood vessel growth. At the new facility, Lilly scientists will also carry out research into immunotoxins for delivery of cytotoxic drugs to their targets.

Arthritis

Lilly's research in this disease is directed towards improved treatment for the inflammation caused by osteoarthritis and also towards ways to slow or stop the progress of the disease.

Immunological research

Another area which will be investigated at the Biomedical Research Centre will be basic immunology and it is hoped that this work will lead to new drugs for autoimmune disease, immunodeficiency syndromes and for immunosuppression for transplant surgery.

In addition to its own in-house R & D, Lilly has also developed links with other companies. The company has an agreement with the NBF Bio-Response for the production of monoclonal antibodies. Eli-Lilly appears to be in a period of transition from a company based on chemical technology to one in which biotechnical processes and molecular biology expertise are of major importance to competitive success. In keeping with other major pharmaceutical companies, Lilly has recognized the key role of basic research for future competitiveness and has taken action to ensure that it is one of

the leaders in the new industry.
Sources: [35–38, 43, 53, 57, 66, 97, 98].

7.7 Monsanto

Origin and development

Monsanto is a diversified chemical corporation with total sales in 1983 of $6.299 billion. The company has businesses in the areas of agricultural products, commodity chemicals and petrochemicals, polymers, resins and plastics and capital equipment and pharmaceuticals. In the agricultural products area (which represents 20 per cent of total sales), Monsanto is the largest international manufacturer and marketer of herbicides; brand names 'Lasso' and 'Roundup' are leading products. The construction and home furnishing market is Monsanto's second largest market with major businesses in industrial coatings, paint resins and plasticizers, nylon fibres and polyvinyl butyral plastic interlayer for laminated glass. Monsanto is also involved in materials and electronic controls for the automobile industry. The Monsanto subsidiary, Fisher Controls International manufactures capital equipment including control valves, field measurement instrumentation and distributed digital control systems for industrial processes. Monsanto created a 'healthcare division' in 1982 to move into the prescription drug market. The firm already supplies pharmaceutical ingredients to producers of aspirin and acetaminophen. Monsanto's involvements in biotechnology have been along three main lines:

- joint research with universities;
- co-operative arrangements with other companies;
- in-house R & D.

Monsanto has made agreements with a number of universities. Its largest agreement is with the medical school of Washington University (St. Louis, Missouri) which will cost $23.5 over five years for joint research on how peptides and proteins regulate cell growth. Two-thirds of the work is in applied projects and one-third in basic research. University researchers can publish the results of the work funded by Monsanto but the firm has the right to prior review and the right to request a short delay if patentable material is involved.

The research projects are selected by a six-person committee with equal representation from Monsanto and the university. Patent rights are retained by the university but Monsanto has exclusive rights to any licenses arising. A number of Monsanto scientists are working in the university under this arrangement. This agreement is somewhat different from some others between universities and industry agreements since it involves a high degree of joint research directed towards certain goals. Other arrangements have involved university support for basic research to gain 'a window' on the new technology and to have corporate scientists trained. In 1982 Monsanto also signed an agreement with Rockefeller University (New York) to carry out basic research in plant molecular biology. The funding is $4m over five years. As in the Washington University agreement, Monsanto allows the academics to publish and has licensing rights on the technology. Another agreement between Monsanto and Oxford University involves an expenditure of $1.8m over five years for research on the role of oligosaccharides in regulating cell function.

The company has also entered into a variety of arrangements with other companies. With Collagen, a company it has equity interests in, Monsanto has a joint development agreement for human bone growth factors using recombinant DNA technology. Monsanto will provide research funding to Collagen under the arrangement and will receive royalties on sales. While Collagen will retain licensing rights to the growth factors, Monsanto is also involved in licensing its own technology to other companies. In September 1984, it was announced that Invitron Corporation had been granted an exclusive license to commercialize large-scale, mammalian cell culture technology developed by Monsanto. This technology has been used within Monsanto since 1983 to produce monoclonal antibodies and to grow mammalian cells involved in recombinant DNA research. Invitron is currently constructing a new production facility (due for completion in mid-1985) and is using the technology to produce substances for pharmaceutical companies on contract. Invitron has an exclusive license to the Monsanto cell culture technology.

In 1982 Monsanto purchased the wheat gene facilities of the seed company DeKalb Ag Research in Kansas and formed a subsidiary, HybriTech Seed International. In 1983, HybriTech Seed International acquired in turn Jacob Hartz Seed Co. which was involved in soyabean development and sale. The acquisition of seed companies and plant biotechnology companies has been one of the main mechanisms employed by large chemical corporations to 'gain a window' on the newer areas of plant biotechnology. (In late 1984 a new biotechnology firm, Agrigenetics was acquired by Lubrizol.) Monsanto's in-house R & D over the past few years has emphasized plant biotechnology and related areas. The firm has developed a Ti plasmid transformation system for genetic engineering in plants and has a number of significant achievements in plant recombinant DNA research. The company is also investigating the use of plant cells as hosts for pharmaceutical production. In 1984 Monsanto announced that it had used the Ti plasmid system to clone and express human chorionic gonadotropin in petunia tissue culture.

Monsanto researchers have developed a genetically engineered bacterial strain (*Pseudomonas*) for use in biological control of pests. The *Pseudomonas* bacterium contains genes coding for a powerful insecticidal toxin derived from *Bacillus thuringiensis*. *Pseudomonas* will be applied as a seed coating on corn to protect it from attack by insect pests. The company has applied to the Environmental Protection Agency (EPA) for approval to field test this product.

The company is also developing methionyl bovine somatostatin by use of recombinant DNA technology and expects to market this in 1990 for the improvement of milk production. In late 1984 Monsanto opened a new 900,000 square feet biological research centre at Chesterfield (Mo.) and expects 600 research staff to be working there by the first quarter of 1985.

The company's current policy is to move away from petrochemicals and concentrate on higher value specialty products in three main areas; traditional chemicals and chemical products, innovate biological products and engineered goods and systems. This strategy is shared by other major chemical companies who see the move towards specialty chemicals and products as a way of dealing with the slow growth of commodity chemicals in recent years. Monsanto's involvement in biotechnology must, therefore, be seen in the context of this overall strategic reorientation. The company has concentrated over the past four years on two areas of biotechnology namely, agricultural and health care. The developments elsewhere in agriculture-related biotechnology such as new pesticides, plant genetics, nitrogen-fixation research etc. posed a potential threat and Monsanto has responded by investing significantly in these areas.

In summary, Monsanto has recognized both the threats and the opportunities presented by biotechnology. In the agricultural products area the company has invested heavily in in-house R & D, company acquisitions and joint arrangements, and it is well positioned to play a leading role in the applications of biotechnology to the agro-food area. The strategic acquisition of gene pools of seed companies and the very strong in-house expertise in plant genetics are likely to give Monsanto a commanding position in a number of new agriculture-related markets. The company has also diversified into prescription pharmaceuticals attracted by the major biotechnology opportunities opening up in this area. The success of this diversification remains to be seen.

Sources: [15–17, 19, 41, 45, 89, 98].

8 The strategies adopted by the companies

8.1 The new biotechnology firms

These firms face the strategic challenge of developing a defensible competitive position whether in products or services. There are a number of important common elements in the strategies that have been adopted by these companies. These include:

- financing;
- links with large corporations;
- forward integration.

The entrepreneurial NBFs have relied on the same sources of finance including private investments by corporations, venture capital, public share offerings and R & D limited partnerships. The extent to which firms rely on the different financing mechanisms has varied. Cetus raised over £100m in its public share offering while Genetics Institute has opted to remain a private corporation. For the leading NBFs there has been a tendency to follow a pattern of financing instruments as follows:

- early development funded by venture capital with one or more rounds of refinancing;
- public share offerings to raise large amounts of cash;
- further development expenses financed through R & D limited partnerships.

On the operational side most NBFs have had to run a deficit for the first few years as they finance R & D and investment in plant.

Their undeveloped company structures have compelled NBFs to enter strategic partnerships with other companies. These have involved joint development projects, licensing of products for manufacture by larger companies, marketing agreements, joint ventures involving a number of functions. The tendency has been for NBFs to evolve in their relationship with other companies so that progressively greater control and payback accrues to the NBF. Thus the stages of this evolution are:

- contract R & D;
- distribution by other companies of products manufactured in-house;
- own manufacture and marketing.

The last stage above indicates one of the critical strategic choices facing the NBFs – how to achieve forward integration. Questions here relate to the desirability, the appropriate product areas and the management of resources. The forward integration from a research services company may be identified as a strategic goal at the early stages or even foundation of the company. Alternatively it may evolve over a period of years. Obviously the choice of such a goal depends on the market segment addressed and the resources available to the company.

8.2 Types of strategy

Despite the above common features, a variety of different strategic choices have been made by the NBFs. These can be analysed by considering four options: 'broad' or 'narrow' degree of market focus and a time-frame for the R & D effort related to 'early' or 'late' products. Taking these factors in matrix form, we can identify three types of NBF strategy (Table 8.1):

- 'focussed' – narrow market focus/late products;
- 'broad-based' – broad market focus/late products;
- 'early products' – narrow market focus/early products.

It is possible to relate these to the three genetic types of strategy which are pursued by firms [104, 105]: overall cost leadership; overall differentiation; focus. The 'focussed' and 'early products' strategies are both types of focus strategy in the Porter definition while 'broad-based' is either overall cost leadership or overall differentiation (see Table 8.1). Thus for the NBFs already considered above in the case studies, Genentech and Genex are characterized by a focussed market strategy and in general R & D programmes with a long lead time, although Genex is ahead of Genentech in forward integration.

Table 8.1 Types of strategy of the NBFs

	Product timing	
	Early	Late
Broad		'Broad-based strategy'
Market focus		
Narrow	Early products strategy	'Focused strategy'

This general strategy is referred to as 'focussed' for short. Cetus and Amgen are also characterized by long lead time R & D but are more broad-based in terms of the markets which they hope to enter. The strategy of this type is referred to herein as 'broad-based'. Finally, those companies which are pursuing Mab applications are involved in short lead time R & D as well as longer lead time R & D and are generally highly focussed. This type of strategy includes companies such as Centocor, Celltech & Hybritech and is referred to here as 'early products strategy'. We will now consider the benefits and risks of each of the different types of strategy.

8.3 'Focussed' strategy

Benefits

Focussing on a specific market niche to obtain overall product differentiation or overall cost leadership is one of the classical forms of competitive strategy. The resources of an NBF, which are severely limited *vis-à-vis* potential competitors in the pharmaceutical industry, are utilized to the maximum extent. This maximization includes the use of research personnel, the optimum amount of contract R & D versus in-house development and the development of production and marketing functions. Having clearly targeted its products in a specific market the firm can manage its arrangements with other companies to fit in with its strategic goals rather than forming opportunistic external arrangements depending on what the R & D programme comes up with.

A clear market focus and a consistent internal organizational structure allow the small firm to capitalize on its technological advantage over larger competitors and to compete on the basis of the most advanced technical products on the market. Some of the firms which are successfully pursuing this strategy include Genentech, Genex and Biogen. In the case of Genentech, the focus is on becoming an integrated market leader in human and animal therapeutics. Genentech has not been tempted to enter diagnostics, but while carrying out contract R & D to provide revenue, has devoted all of its efforts to long-term research into therapeutics. These are low volume high-value products and the strategic time frame of the company extends into the 1990s when it intends to be the industry leader in a wide range of new drugs based on recombinant DNA technology. Genex, on the other hand, has targeted a very different market, specialty chemicals, but the degree of market focus is equally evident. Genex is focussing on areas such as industrial enzymes, amino acids, vitamins, etc. which are already products today. The strategy is to achieve cost leadership *and* overall product differentiation within these selected niches. The lead time has been long and the company has sustained substantial operating losses in recent years in pursuit of its strategy. However, the benefits here should be a strong competitive position relative to existing specialty chemical producers and also any potential new competitors.

Risks of focussed strategy

There are also considerable risks associated with a 'focussed strategy'. For the leading NBFs the risks are that they may be 'out-focussed' by the major established corporations. Once the technical and commercial feasibility of say a new therapeutic has been established by a pioneering NBF, which has taken most of the risk, an established corporation may devote larger resources to that product or one closely similar and so achieve lower costs. Pharmaceutical competition however, is generally non-price based and so superior product performance may still imply large market share. But the established corporations will also be competing on product performance and may be able to introduce second or third generation products at a more rapid rate than a NBF.

For the smaller NBFs the risks are both of a different nature and perhaps more intense. The firm which has focused on a specific market niche may find a dislocation between its present and expected resources and the market it is aiming at. Thus for small NBFs in the human health care area, there may be considerable difficulties in raising sufficient funding to achieve forward integration. Too much focus on an area in which the company cannot realistically expect to build an integrated business is obviously a strategic mistake unless the company wishes to become a specialized service company. Too narrow a focus is also undesirable. Companies such as Interferon and Endorphin are focussed on a few very specific products and in a rapidly changing scientific and technical environment this type of overspecialization is extremely risky.

8.4 'Broad-based'

Benefits

This type of strategy is characteristic of NBFs such as the early Cetus and Amgen. The benefit for a research company in adopting this approach are the reduction to some extent of competitive risk in particular areas by having a broad R & D portfolio. The objective here is to achieve product differentiation through the marriage of the company's superior technology and appropriate joint ventures in a range of products or industries.

Many authors of R & D studies accept the concept of economies of scale in R & D. If this is accepted then, a broad R & D portfolio of the type described above would allow maximum utilization and return on R & D effort in an NBF. The same techniques of recombinant DNA, enzymology, etc., which are applicable to development of biotechnology products in one industry may be used also for product development in a range of other industries. This strategy is aimed not only at risk reduction through diversified R & D, but also at gaining industrial leadership in a range of biotechnology products through the commercial leverage supplied by established companies. The NBF does not become a totally integrated company in each of the markets where it is competing and so avoids the risks this would entail.

Risks of broad-based strategy

There are also considerable risks associated with this strategy. The critical question mark here relates to the ability of the broad-based NBF to compete with a focussed strategy either from another NBF or a large established company. By spreading its resources, both R & D, production, marketing and managerial over a range of areas, the NBF risks not having sufficient resources in any area to compete with focussed NBFs who are devoting all their effort to these particular areas. Furthermore, an established firm may 'focus' on any of the markets being addressed by the broad-based NBF and out compete it. It is likely that Cetus, for instance, realized the dangers inherent in too broad-based an approach and made considerable efforts to narrow its focus.

The relationships with other companies inherent in this strategy also have associated costs. The exploitation of the NBF's research achievements through joint ventures and other strategic partnerships means that the company is not using its proprietary technology to the full extent for purposes of forward integration. This places the company in a position where it risks becoming a specialized service or technology company as other NBFs enter full-scale manufacturing and marketing. Furthermore, the transfer of technology to joint partnerships risks the diffusion of the technology to the other partner and also, where the NBF has a minority interest, a possible loss of control over how and when such technology is used. The broad-based strategy is regarded here as being the most risky for an NBF. It is perhaps more suitable for a broad-based multinational corporation than for an NBF.

8.5 'Early-products' strategy

Benefits

This is the strategy which has been adopted by those companies concentrating on monoclonal antibodies (Mabs). It can be regarded as a type of 'focus strategy', although it differs from the 'focussed' strategy described above in having a short product development time because of the nature of the area on which it concentrates. The strategy is rapidly to introduce a range of Mab products (Mab diagnostics) in an area where entry barriers are low. This generates short-term cash flow which assists with the longer term health care applications being pursued. Examples include Celltech which introduced its first products in one year after start-up and Centocor which introduced three important new assays, three years after start-up. The advantages of this are fairly obvious. The area of Mab diagnostics is the most rapidly growing of the biotechnology industry and the NBFs capitalize on this. Other benefits here are the same as those described above for the 'focussed' strategy.

Risks

The field of Mab diagnostics is already becoming very crowded. Lower entry barriers imply intense competition and the major established companies are responsible for many of the new products already introduced in this market. The NBFs are and will be competing in the marketplace with established companies which in many instances already have established businesses in non-Mab diagnostics. In addition, recent changes in the health care reimbursement policy in the United States are having an adverse effect on the diagnostics market.

Biotechnology firms which are concentrating on Mabs such as Celltech Centocor, Hybritech, and Monoclonal Antibodies face important strategic decisions relating to early versus late products. If a high proportion of the product development effort is devoted to early products then the NBF risks losing out on the race for much more valuable products, such as *in vivo* imaging, that take longer to develop. Given the intensity of competition in the area of *in vitro* diagnostics it is probably wiser for the NBF involved in Mabs not to invest too much of its R & D effort in this area but to concentrate on using its proprietary Mab technology for longer term products.

8.6 The established firms

Response to technological change

Established companies faced with a rapidly growing new technology have a variety of options including:

- do nothing;
- monitor only;

- attempt to prevent the development of the new technology:
- improve the old technology:
- participate in some manner.

In the latter half of the 1970s the established American and European pharmaceutical companies found themselves confronted with the rapid development of the 'new biotechnology'. These companies had no in-house expertise in areas such as genetic engineering and the development of the NBFs represented a threat either short or long term to their existing pharmaceutical business. Given the early level of interest in interferon, companies involved in anti-cancer drugs felt particularly vulnerable to technological obsolescence. The established pharmaceutical and chemical companies have responded to this threat with a remarkable degree of flexibility and have participated in a variety of ways. The forms of participation have included:

- equity investments in NBFs:
- joint ventures and licensing with NBFs:
- in-house R & D investment:
- investment in academic institutions.

Many of the leading companies have taken equity in a number of NBFs. Koppers for instance has equity in Genex, Engenics and DNA Plant Technologies. Johnson & Johnson has taken equity in three companies concerned with diagnostics and/or Mabs including Quadroma, Enzo Biochem and Immulok. Schering-Plough has invested in Biogen and also acquired DNAX. It is interesting that it is not only pharmaceutical companies which have invested in the NBFs: chemical companies such as Monsanto, engineering corporations as represented by Fluor and energy and oil corporations (Phillips, Petroleum, Standard Oil of California) have also been involved.

Most of the interaction between the NBFs and the established firms has been in the area of product licensing by the NBFs. For the established corporations this represents a quick way of catching up technically and competing in early product markets while avoiding the long lead times experienced by the NBFs in the development of their products. Established companies have also engaged in joint ventures with NBFs such as that between W.R. Grace & Cetus (Agracetus) and Genencor, a joint venture between Genentech and Corning Glass.

Established corporations have invested heavily in building up in-house R & D in recombinant DNA, immunology and other new techniques. In 1984, Du Pont opened a new life science R & D complex which cost $85m. Eli-Lilly has built a $50m biomedical research centre and other companies are also making very sizeable investments. Because the original source of the new technology was academic laboratories, established companies have paid particular attention to funding academic research and to linking their own research programmes with those of academic institutions. The aim here is to engage in technology transfer activities from the university laboratories to the companies and also to have a 'window' on new basic science which will be useful in the future.

In summary, the established companies have invested to access technology and early products and in areas where development is longer. We shall now consider how their investments contribute to competitive strategy and the strategic choices facing the established firms.

8.7 Technological leadership and followership

The NBFs are by their nature pioneering and technologically advanced. The established companies are late entrants into the biotechnology industry. Thus a crucial choice in technological strategy for the established companies is between technological leadership and technological followership. It might be thought that this choice was directly related to the timing of entry. i.e. pioneering being equivalent to technological leadership and late entry implying technological followership. However, the situation is more complex than this. According to Porter [104, 105], there are three options: pioneering technological leadership, late-entry leadership; and late-entry followership. In late-entry technological leadership the company does not attempt to gain from the experience of the leader but instead attempts to 'leapfrog' forward to the second generation technology. This is a particularly important strategy in the context of biotechnology as we shall see below. The choice between technological leadership and technological followership is determined by a range of industry structural characteristics. These are listed by Porter as

the technological opportunity to influence cost or differentiation.
the uniqueness of the firm's technological skills, first mover

advantages, the continuity of technological change, the rate of change in process technology or customer purchasing behaviour, the irreversibility of investments, uncertainty and leadership externalities [105].

Before examining the application of these factors to biotechnology in a theoretical sense, it is necessary to consider how the established companies have behaved. Established company behaviour is very complex because of their diversification in general and the wide range of biotechnology markets which are opening up. In generic strategy types, an established company may focus in one market, and attempt lost leadership or differentiation in others. Similarly, in technological strategy, which can be used to achieve any of the above goals, the same company may pursue both leadership and followership, pioneering and late entry. Some examples of this are given below.

Eli-Lilly

The strategy being pursued by Eli-Lilly has been described above in relation to human insulin and other developments. Lilly's investment in human insulin (achieved through technology licensing) was an example of pioneering technological leadership. Lilly owns the world's first large-scale plants for recombinant DNA human insulin and the company hoped to capitalize on first mover advantages and the technological opportunity provided to achieve major product differentiation. Like many other established companies, Lilly is not relying exclusively on first generation biotechnology products and is investing heavily on in-house R & D. This can be regarded as a policy of late-entry technological leadership. The investment in long lead time R & D is expected to result in second generation products which will leapfrog the technology of the present and new future. This is a policy being followed by most multinationals in addition to their other activities in specific markets.

Imperial Chemical Industries

Imperial Chemical Industries (ICI) invested in the development of single cell protein (trade name 'Pruteen') and commissioned their first and to date only plant in 1980 after many years of R & D and an estimated investment of $150m. This is another example of pioneering technological leadership by a major established company. The plant contains the largest sterile fermentation unit in the world and involved considerable engineering innovation. The product 'Pruteen' has, however, been unable to compete with soya on price. Despite this setback ICI are also pursuing a policy of late-entry technological leadership through their investment in academic research. The company is currently investing in a number of university centres. One especially interesting example is the laboratory at Leicester University (United Kingdom) which ICI funds to carry out research into cloning in mammalian cells, a production system which might, in the future, prove more useful than culture of genetically engineered micro-organisms.

Upjohn

Upjohn consolidated its in-house biotechnology R & D into one unit in 1983 and expects to build this up over the next few years. The unit will carry out biotechnological R & D in support of Upjohn's other research activities in human health care and agriculture. The therapeutics units, for instance, will determine what substances are needed and the biotechnology unit will then try to produce these substances. This would appear to be a policy of technological followership. The company is not rushing to pioneer first generation products but is holding back until the market situation becomes clearer.

W.R. Grace

W.R. Grace is pursuing a number of developments: its investments include basic and applied research at university centres and acquisition of companies. The company is reportedly working to a twenty-year time horizon in its biotechnology strategy which is aimed both at protecting existing areas of business such as nitrogen-based fertilizers and also at expansion into new areas of health care and specialty chemicals. The strategy here also appears to lean towards technological followership. W.R. Grace has not developed its own products or licensed products for competition in the near term. It is building up a strong internal capability and will participate in new markets when it can benefit from the experience of technological leaders.

Summary

The technological strategies of the established companies include all three options. Some companies such as Eli-Lilly and Schering-Plough have used the mechanism of licensing to acquire technology and so engage in pioneering technological leadership. Many other established companies are pursuing a policy of technological followership in which they refrain from entering various markets until uncertainty is reduced but meanwhile build up in-house expertize and engage in arrangements with NBFs such as equity participation. Most of the established companies are investing heavily in in-house R & D and in sponsoring research in the universities. This can be regarded as a preparation for late-entry technological leadership. While the established companies cannot be said to be pursuing this strategy at present, it would appear that they are positioning themselves technologically to pursue such a leapfrog strategy to second generation products.

8.8 Strategic options: NBFs and established firms

As indicated above the choice between technological followership and technological leadership is central to the policies of the established companies. The NBFs do not have the same degree of freedom. Small high technology companies without significant product sales must in general engage in pioneering activity. It is useful to consider here the factors contributing to the choice of technological strategy and which strategies are theoretically favoured in the biotechnology industry.

Technological opportunity

This expresses the 'technological gap', representing, for example, a unique product or low-cost process, which a company can place between itself and its competitors. Biotechnology is characterized by a greatly accelerated increase in new knowledge. Major opportunities exist, therefore, for firms to achieve significant technological innovations and maintain a defensible position through proprietory protection of their new knowledge. It is because many firms, individuals etc. have perceived this technological opportunity that biotechnology has generated so much entrepreneurial activity and public interest. Emphasis on this factor implies a policy of technological leadership.

Uniqueness of the firms' skills

This factor undoubtedly favoured technological leadership in electronics. However, it is not relevant in biotechnology. Because of the wide diffusion of biological and bioprocessing knowledge and the absence of enforceable patents in the 1970s, biotechnological skills are widely diffused. Hundreds of firms in the United States and many others world-wide have access to the technology. There is also a spectrum of alternative technological approaches to common problems which lessens the uniqueness of a firm that has excellence in one of these approaches. Hence, the lack of technological uniqueness here favours technological followership.

First mover advantages

These are advantages arising from moving into a particular field first, other than technological lead, such as reputation, switching costs to alternative products, etc. Being first in the pharmaceutical market is generally regarded as a major advantage and it is not unheard of for a company to abandon a development when beaten in this race [110]. For biotechnological pharmaceuticals and other biotechnology products there are also likely to be first mover advantages. Getting an innovative product to market first may mean the ability to satisfy a large amount of demand and achieve dominant market share while competitors are still at the developmental stage. Reputation is an important factor in biotechnology and if a company can establish a 'leading edge' image by pioneering new products this can lead to advantages over competitors in terms of customer behaviour. This factor would support a strategy of technological leadership.

Irreversibility of investment

A high degree of irreversibility of investment favours technological followership. In the biotechnology industry there are a wide variety of production technologies depending on the type of industry. The production of a diagnostic kit cannot be easily grouped with micropropagation of plant varieties. However, leaving aside the agricultural applications, the processes of fermentation or cell culture are central to many biotechnology businesses including monoclonal-based diagnostics or specialty chemicals or pharmaceuticals. Each product produced by fermentation or cell culture requires specific tailoring or optimization of the process including the recovery stage. However, the investment in the plant itself is flexible since the same plant can produce a wide variety of substances depending on substrate and micro-organism employed and the same unit processes can be used for recovery of a range of substances. This would seem to suggest that investment in fixed assets is not irreversible.

However, in the manufacture of substances using recombinant DNA technology there are factors suggesting irreversibility. Each recombinant DNA product requires a specific expression vector for the gene which encodes the product. Hence each recombinant DNA product requires a specific process at the molecular level, although the gross aspects of the equipment and process will be similar to other products. Because a number of host organisms and expression systems can be used in the production of any substance, there is not a unique one-to-one relationship between recombinant DNA product and process but each process must be designated as it were to making only one product. The patentibility of expression vector systems has implications for switching products. A firm wishing to switch products may find that it has to employ one or more proprietary processes. These may be either unavailable or available only with high royalty payments, producing a large cost disadvantage in switching. Therefore, although fermentation and cell culture technology is itself flexible there is some irreversibility involved in processes at the molecular level. This factor does not unambiguously favour either leadership or followership.

Technological uncertainty

This is closely related to irreversibility of investment. A high degree of uncertainty exists about products and processes in the emerging biotechnology industry, which is reflected in the risk inherent in investing both in R & D programmes and in manufacturing plant. Uncertainty also exists over the demand for various products. Therapeutics in particular are being introduced in an environment where new knowledge on their function and that of rival products is being continuously generated. This uncertainty would appear, therefore, to favour a strategy of technological followership.

Leadership externalities

These are the extent that technological followers get a free ride on the investment by the leader. They include such factors as gaining regulatory approval for product and processes, customer education and marketing costs. Biotechnology products will be regulated on a case-by-case basis. Although pioneering products may experience some delay due to the need to educate regulatory officials, the regulatory framework is such that followers as well as leaders will incur substantially the same delays and costs. For instance, drugs manufactured by recombinant DNA are regarded as new drugs by the FDA even when identical to drugs already on the market.

Also in a situation of very rapid technological change, investment by a leader in 'externalities' will not necessarily benefit a follower since the product may be obsolete by the time the follower decides to invest. This would favour technological leadership. This analysis of strategic options is summarized in Table 8.2.

Therefore, of the six factors, three favour technological leadership (technological opportunity, first mover advantages and leadership

Table 8.2 Factors affecting choice of technological strategy in biotechnology

Factor	Status as regards biotechnology	Favours
1. Technological opportunity	High	Leadership
2. Uniqueness of firm's skills	Low	Followership
3. First mover advantages	High	Leadership
4. Irreversibility of investment	?	Uncertain
5. Technological uncertainty	High	Followership
6. Leadership externalities	Low	Leadership

externalities), two favour followership (uniqueness of skills and technological uncertainty), and one (irreversibility of investment) is uncertain. The essential question here is the relative importance of the various factors. On balance it would appear that technological leadership is favoured. Because of the multitude of potential applications of the new technology and its rate of change major opportunities are opening up for companies to create a differentiation gap between themselves and their competitors in various markets. The strategy of late-entry technological leadership or 'leapfrog' is particularly important here since it would allow the established firms to jump ahead of the pioneering NBFs. Technological followership may also be a wise strategy in those markets where there is a very high level of product and process uncertainty and this would appear to be the implied strategy of some established firms.

The licensing arrangements between NBFs and established companies must be regarded as being more beneficial to the established companies than to the NBFs. Because they lack forward integration, NBFs have to license their technology to companies which are likely to be their competitors in the future. This involves a transfer of technology and a possible closing of the gap between the technological leader (NBF) and the established company (which may also be pursuing a technological leadership strategy but through licensing and acquisition rather than R & D). For the established company such a course taps a cheap source of technology. Thus established companies would appear to have considerable competitive advantages over NBFs. The established companies have more strategic 'degrees of freedom' than the NBFs. Their large resources and diversified business portfolios allow them to pursue a range of different but complementary strategies across their product ranges. They can access the technology in a large variety of ways including licensing, joint ventures, sponsorship of academic research and in-house R & D and they have the distribution channels for global competition.

This leads to some tentative suggestions about the future state of the biotechnology industry. The industry structure will be shaped by those companies which can establish defensible niches, namely, the established companies and a small number of the NBFs. The majority of NBFs will be unable to influence industry structure due to intense competition from the industry leaders. The established companies are likely in many instances to leapfrog ahead of NBFs into more successful second generation products. The fate of the majority of NBFs, therefore, is likely to be one of acquisition, cessation of activity, or development as specialized service companies which have maintained their technical lead.

9 The elements of success

Success in the biotechnology industry will depend on a number of factors including the nature of the industry within which the biotechnology product is introduced. Each of the industries affected by biotechnology has certain key success factors. Thus for bulk products such as single cell protein or commodity chemicals low unit cost and access to cheap raw materials are essential, while for pharmaceuticals innovative R & D resulting in new products are equally essential. In addition, there are a number of common factors affecting the success of those companies pursuing biotechnology innovation. These are related to the nature of the company and its positioning, to the emerging nature of the industry and to how strategic decisions are taken in an environment of uncertainty. Some of the most important factors necessary for competitive success are identified and discussed below. The role of management and its reaction to uncertainty factors will be crucially important.

9.1 Strategic management

The techniques of corporate planning have become institutionalized within international business over the last ten to fifteen years. Recently there has been some criticism of the over-reliance on the quantitative methods of planning and a movement 'back to the basics'. Quantitative methods based on, for example, return on investment, portfolio planning and use of the life-cycle concept have all come under attack and Hayes and Abernathy in an influential paper in the Harvard Business Review have stated that American management practices based on 'analytic detachment' rather than hands-on experience and short-term cost reduction rather than long-term technological competitiveness have played a major role in undermining the vigour of American industry [72]. This debate goes beyond the scope of this book, but it is evident that a new synthesis of strategic management is emerging in which the planning techniques complement rather than dominate basic factors such as 'hands-on experience', company culture, motivation of the workforce and responsiveness to customer needs. A wide discussion of strategic management techniques is impossible here. However, three important techniques, namely portfolio planning, the experience curve concept and technological forecasting are discussed below as they have a particular bearing on the strategic constraints and opportunities facing the established biotechnology companies and the NBFs.

Portfolio planning

There are a variety of portfolio techniques most of which result from the 'growth/share matrix' developed by the Boston Consulting Group. Most companies, except the simplest, consist of a variety of businesses. These businesses can face different strategic options depending on how they are placed in terms of growth and relative competitive position. Growth rate and market share are presented as a matrix in the Boston Group analysis (see Table 9.1).

Table 9.1 The growth share matrix of the Boston Consulting Group

	Market share	
	high	low
Growth rate high	stars	question marks
low	cash cows	dogs

Individual business may be classified into four types on this basis, high growth/high share (stars), low growth/high share (cash cows), low growth/low share (dogs) and question marks which are high growth/low share. Stars present the best investment opportunity but require large investments of money to maintain position. As stars mature they turn into cash cows, i.e., businesses with low growth and hence low reinvestment needs but high generation of profits. Cash cows are used to provide surplus cash for investment elsewhere in the business. Question marks absorb large amounts of cash because of their growth but generate little due to their low share. They may go in either of two directions becoming stars if managed correctly or alternatively sinking to become dogs. Dogs with a low market share and low growth rate are a total contrast to stars and can become cash traps perpetually sucking in funds without any possibility of achieving a viable cost position.

According to the theory cash generated from the cash cows is used to consolidate stars which are not self-sustaining and to convert a select number of question marks into stars. The matrix can be quantified for an analysis of an actual company's portfolio. Such a matrix is very useful in the planning of biotechnology developments within the established companies. Like mature companies, the established biotechnology companies have businesses in each quadrant. Most importantly they have cash cows which are generating large profits. In the pharmaceutical industry such businesses may be products such as antibiotics which are nearing the end of the patent life but are well established on prescription lists and generate large profits. New biotechnology-based drugs will hopefully become stars and market share can be built using the finance available from the older established drugs. In other words, biotechnology products can be regarded as a separate type of business in the portfolio matrix. As the first generation of biotechnology products age and become cash cows, the established company can redefine its biotechnology businesses as a separate portfolio, i.e., the profits generated from the biotechnology drug could be used to consolidate a new biotechnology business in agricultural products, for example.

For the NBF, portfolio planning is less useful since they do not have any cash cows and possibly no stars either. Their initial products will probably be question marks and they will have to use their available equity cash to convert them to stars. One group of NBFs is attempting to accelerate the development of stars and cash cows within their portfolio and this is the group of companies pursing an 'early products' strategy, described above.

The experience curve

The Boston consulting Group also developed the use of the experience curve. The experience curve expresses the relationship between the total cost of manufacturing and distributing a product and the cumulative production volume. It has been observed that each time the accumulated experience in manufacturing a product doubles, the cost in real terms (i.e., in 'constant money') also declines by a characteristic percentage generally between 20 and 30 per cent. [116]. The decline in unit costs with experience is caused by a variety of reasons including capital/labour substitution, economies of scale and increased worker efficiency over time. This may be a useful tool for strategy development. There are two key strategic considerations arising from this technique:

- an uncompetitive cost position will result from a failure to reduce costs along an appropriate experience curve;
- failure to grow as rapidly as competitor companies will also lead to an uncompetitive cost position.

The largest competitor will have the potential for the lowest unit costs and therefore highest profits while the smaller competitor will be unprofitable unless he can devise a strategy for obtaining a dominant share either overall or in a particular segment [30, 116]. Thus this concept should be particularly useful in analysing competition between NBFs and established firms and between small and large NBFs. The experience curve concept is particularly important in relation to the question of market focus and this is discussed further below.

Technological forecasting

Technological forecasting comprises a wide variety of techniques, both quantitative and subjective which can be used to attempt predictions of future technological trends and states [76]. These techniques range from quantitative extrapolative methods to scenario writing, and from subjective assessments of experts as in Delphi to cross impact analysis. Successful forecasting is more an 'art' than a 'science' and over-reliance on quantitative methods may lead to over-confidence in the precision of the forecasts. Technological forecasting is widely employed today in most large corporations and a range of organizational structures for this function have evolved. A recent study of such activity in American corporations found that forecasts, although generally made by a specific group within the company, are likely to be more effective if the staff is operating in a cross-functional matrix regardless of their administrative home and if they are supported by and involve top management, line management and staff in many parts of the organization [84]. Technological forecasting is particularly important to the R & D function and as biotechnology is R & D intensive it is highly relevant here. It can be used in the emerging biotechnology industry in the selection of the R & D portfolio, and most importantly at ech stage of the resource allocation process. This is discussed below under the management of uncertainty.

Limits of management

Different biotechnology companies will have access to different types of management expertise. The established companies in a sector such as pharmaceuticals have a very strong base of expertise and experience in production, marketing, regulatory affairs and clinical trials, R & D and strategic planning. Many of the NBFs, on the other hand, were founded by entrepreneurial scientists and experienced line managers were brought in later from industries such as the pharmaceutical industry. There is no shortage of such mobile expertise and many NBFs have been able to build up excellent management teams. The quality of management expertise will obviously be one of the crucial determinants of success in the biotechnology industry. But its importance must not be overstated. A recent book on the subject of corporate success [102] appears to place excessive emphasis on management excellence. The view here is that managerial expertise is only one of a number of variables contributing to success including access to technology, financing and industry positioning.

9.2 Management of uncertainty

Given a rapid rate of technological change and high uncertainty, a high degree of flexibility is required in the biotechnology business. This flexibility is particularly important in resource allocation and the acquisition of technology. Within the R & D process a number of stages can be identified from basic or applications-oriented science through applied research, scale-up and commercialization. Each stage involves the allocation of resources in an uncertain environment. To minimize risk, it will be necessary to review the environment and evaluate resource allocation decisions in this context. This should be performed at each stage in resource allocation. In practice this should involve a formal technological and environmental scanning function within the company and the integration of this into strategic management so that every decision on resource allocation is informed by this information to the greatest extent possible.

The sources of information which the company can use include:

- scientific research findings (published);
- unpublished research results from personal contact or funded externally by the company;
- published or unpublished information on activities of competitors;
- external non-technical factors such as government regulation, legal precedents and changes in health care reimbursement systems.

When resources are allocated to the development and commercialization of a particular product, it is necessary to maintain maximum flexibility dependent on the environmental scanning process. This involves the establishment of contingency planning for alternative products which could be substituted if the product under development appears unattractive in the light of new information, e.g., a new technical breakthrough rendering the present product obsolete or disappointing clinical trial results.

Also competing products under development by competitors must be evaluated to determine the optimum route for switching to these products should that prove necessary. This can prove highly complex as it may involve a combination of in-house development and external technology acquisition of processes through licensing. It may also be valuable for a company to develop its own process technology for potentially competing products as a type of insurance against the need for sudden switching of product development. Patents can then be taken out to protect such processes in case they are required but at the risk of alerting competitors to the strategic alternatives being considered.

Strategic technological decisions in such areas as licensing technology, acquiring small high-technology companies, in-house R & D and external sponsorship of research in universities should all be informed by the output from the technological forecasting process. As well as influencing the allocation of resources to in-house R & D, technological forecasting can be used to manage an external portfolio based in universities. Continuous monitoring of trends should lead to a continuous reallocation of resources between various long-term projects based in universities as new information accumulates on the probabilities of success in various areas. An effective system must be established within the company for staff participation in forecasting and for dissemination of its results to key policy makers. The main factor in the success of such a process appears to be the attitude of top management. If top management takes an enthusiastic attitude towards and actively directs the forecasting it can contribute significantly to plotting a course for the company. If not, forecasting tends to become a dispensible activity which can be reduced or eliminated in difficult circumstances.

9.3 Market focus & positioning

The degree to which firms focus on specific markets and their relative competitive positions will be important determinants of success. Again there are different strategic choices facing the established firms and the NBFs.

For the established firms the strategic questions relate to the degree to which biotechnology businesses should be incorporated within their portfolios. This must be established initially by carrying out a SWOT analysis (strengths, weaknesses, opportunities and threats). For the established firm faced with the phenomenon of rapid growth of technologically advanced start-ups, the first objective must be to invest in R & D or products so as to avoid risks to its current portfolio. These risks may be both short-term and long-term. The short-term risks may necessitate immediate acquisition of advanced technology through licensing or company acquisition while the long-term risks can be faced through realignment of R & D programmes.

The major opportunities facing the established firms arise from the fact that biotechnological innovation is occurring within markets where they have a dominant position. Their experience in addressing these markets combined with their substantial resources thus make biotechnology a highly attractive group of new markets. Because biotechnology is so transectoral and diverse, not even a diversified multinational corporation can address all its markets equally well. Hence a biotechnology portfolio must be selected. Within this portfolio, alternative strategies are available as described above, i.e. focus, overall differentiation or overall cost leadership and the strategy chosen for each market will reflect the key success factors for that market and the company's ability to compete broadly in that market. For instance, a health care company without previous experience in diagnostics may decide to focus on a specific group of assays, e.g. cancer diagnosis, as a means of obtaining entry into the diagnostics market. The aim here would be to dominate this particular market segment and use this as a basis for further expansion into diagnostics. Alternatively, a health care company with a dominant market position in diagnostics will aim for new technologies which are broadly applicable both to existing and new assays. Unless it wishes to lose market share, it must pursue a strategy of overall differentiation in this market.

There are two extreme positions for selection of a biotechnology portfolio: building on a core business or radical diversification. A company may concentrate its activity in biotechnology on innovations within markets which it already addresses. An example of this type of activity is the launch of recombinant DNA human insulin by Eli-Lilly, a company which dominates the insulin market in the United States (but not in Europe). Alternatively, a company may diversify into new markets within which it has no previous experience. An example of this type of activity has been the investment of a number of Japanese food companies such as Kirin Breweries in recombinant DNA pharmaceuticals. In practice most companies will develop a biotechnology portfolio which is a mixture of new diversification and innovation within existing markets. Diversification by the established companies may not simply reflect the perceived opportunities offered by biotechnology but may involve a long-term strategic re-orientation by the company. Monsanto, for instance, has a strategy of moving away from petrochemicals (20 per cent of net 1983 sales were in polymers) and into a group of businesses involving more high technology: advanced agricultural products (plants and chemicals), human health care and materials science. Biotechnology, therefore, is merely one of a group of key technologies which the company needs to access in order to carry out such a transformation.

Because of the diverse nature of biotechnology products it is difficult to make general statements on the relationship between relative competitive position and success. Relative market share can depend on how the market is defined. Thus in pharmaceuticals, a firm may have a low share of a particular therapeutic group but a large share of a market segment within it. All products within the same therapeutic class are not potentially substitutable. Slatter [110] provides an example of this in his description of the British tranquilizer market. In that market, 81 per cent of prescriptions for one company's product are for treating neurosis and only 2 per cent for psychosis. For a second company, however, 26 per cent of all prescriptions for its product are for psychosis and only 39 per cent for neuroses. Since psychosis tends to be treated in hospital 'the second company's product tends to be a hospital as opposed to a general practitioner product'.

In the emerging area of new anti-cancer therapeutics a wide range of therapeutic possibilities are being investigated including interferons, other lymphokines, tumour necrosis factor, immunotoxins and applications of oncogene research. It is quite likely that market segmentation, based on unique performance characteristics, will emerge here. Thus a company could have a low market share in cancer therapeutics broadly defined but dominance in a particular segment. Hence definition of the market is crucial in any analysis of relative market share.

The experience curve concept must be used with caution in analysing relative competitiveness in biotechnology. If two firms manufacture and compete in a substance and use the same technology, cumulative volume will undoubtedly lead to cost reduction, so that after a period of time one company enjoys a cost advantage. However, because of the very rapid rate of innovation the other company may leapfrog forward to a new experience curve with a radically different cost structure. In a rapidly changing technical environment it would appear that greatest market share, and hence cumulative volume, is no indicator either of a continuing cost advantage or of competitive success. Companies should, therefore, not concentrate on moving down a particular experience curve, but on the generation of discontinuities into the experience curve, i.e., continuous process innovations of a major kind leading to cost reduction unobtainable on the previous experience curve.

Dominance of one or more segments is, however, critical for success in the pharmaceutical business [110]. For the established company, however, a low market share can be tolerated in new biotechnology products since its overall business is not dependent on biotechnology. Income from existing cash cows can be used over a longer time period to build market share or alternatively the market can be abandoned. Exit may be costly but need not necessarily seriously damage the company.

For the NBF which is attempting forward integration, the selection of its portfolio is determined by its proprietary technology and perceived opportunity. While some NBFs have attempted a broad-based portfolio many others have focussed on specific niches. The view here is that a strategy of focussing is the only viable option for an NBF which wishes to be successful in manufacturing. A start-up which does not have any existing products cannot hope to compete effectively in a number of markets or even broadly within one. It must attempt to identify segments within a specific market where its technological lead can be used to comparative advantage. One of the risks of a technology such as recombinant DNA is that it offers very attractive possibilities in many areas. Given an in-house expertise in this technology, a firm may feel tempted to capitalize on its wide applicability. As shown above, this can lead to the significant danger of being out-focussed by other companies. For success, an NBF must clearly define its core business and stick to it. For NBFs concentrating on pharmaceutical applications this means that dominance must be achieved in some segments. Unlike the established firm, the business of an NBF is biotechnology by definition and if it fails to secure a significant market share in those few biotechnology drugs which it has targeted, it may fail as a company. For the NBF involved in specialty chemicals, there may be many products within its portfolio due to the fragmentation of the specialty chemicals market. Therefore, unlike the pharmaceutical NBF, it does not rely on a few major success products.

Some NBFs may not plan (or may be unable) to become integrated manufacturing companies and these companies may become specialized research service companies. Such a company would provide contract R & D and/or develop products and services for external licensing. The 'core business' would involve expertise in particular technologies which could be applied to a variety of sectors or to a narrow area, e.g., services in animal reproduction as offered by Genetic Engineering. Success in this business would depend very much on maintaining unique technological skills such that customers would be at an advantage in using the company's skills rather than developing their own. Some possible opportunities here include protein engineering services, genetic screening, disease susceptibility based on molecular data, specialized software for biotechnology and so on. Market share is not a relevant consideration for this type of business where success depends on the rate of technological diffusion. A high rate of diffusion would allow clients to obviate the need for its services.

9.4 Innovative culture

High technology companies, like other organizations, are not impersonal structures but are shaped by their people and the environment within which they work. A growing awareness of this elementary fact has led to a number of popular management books, which criticize American executives for over-reliance on quantified methods of corporate strategy and for neglecting basics such as closeness to the needs of the customers and the way the workforce is treated. Pascale and Athos [101] suggest that the bond produced by shared values, is

> probably the most under publicised secret weapon of great companies.

They also argue for

> the importance of superordinate goals, the significant meanings or guiding concepts that an organization imbues in its members.

Peters and Waterman in *In Search of Excellence. Lessons from America's Best-Run Companies* [102] also emphasize 'the basics' and state that companies do not succeed through strategic planning alone but through such factors as

> sticking to what they know best, quick action, customer interaction, respect for employees and giving meaning to their mission [87].

These ideas are highly relevant to any consideration of the biotechnology industry and especially the NBFs. In common with other high technology companies in areas such as electronics, NBFs have a corporate culture and set of shared values which is highly conducive to innovation. A large, if not major proportion of the NBF employees are R & D personnel and there are certain common values and environments throughout the various companies. These firms are pioneers and their research personnel (and managers) are acutely aware of both the risks and the major opportunities. This generates a high degree of competitiveness *vis-à-vis* other companies and a strong internal cohesion.

Many of the research workers in the NBFs have moved into commercial life from academia, attracted by the potential rewards which the industry promises. These scientists take with them, a great enthusiasm and excitement about molecular biology. It is a highly stimulating intellectual area and because basic and applied research borders are blurred, as described above, it has been possible for company research scientists to maintain in the NBF, their intellectual interests and enthusiasm rather than feeling that they are merely serving corporate interests. Such scientists are, of course, serving corporate interests but are simultaneously participating in exciting scientific discoveries.

Within NBFs, scientists work in an environment which combines many of the best elements of university research with those of commercial research. The company encourages, in many instances, scientists to work on their own ideas as well as pursuing the company R & D programme. Many companies permit their research staff to publish important findings (applications-oriented science) in leading journals such as *Nature, Science, Cell,* and the *Journal of Molecular Biology*. This allows leading individuals to enhance their reputations within the broader scientific community, both company- and university-based. Reputation is a particularly important form of reward for research scientists and the opportunity to pursue this type of reward within a company structure is highly attractive. For younger research staff the fact that pioneering scientists with established reputations in science are working within the company is also an inducement to make a career in such an environment. In fact, the reputation of the NBF is also very valuable in attracting talented researchers who wish to be associated with 'leading-edge' companies.

In addition, many NBFs are physically located near university centres. In California, Cetus, Genentech and others are located near the Bay area with important universities such as Stanford nearby. Around Boston, there is also a group of NBFs in relatively close proximity to institutions such as Massachusetts Institute of Technology and Harvard University. Such physical closeness permits interaction between NBF staff and university-based researchers. A number of NBFs also have stock option plans for company members.

The organizational structure of R & D within NBFs is also similar in some respects to university research teams. Instead of being focussed around specific products, research teams in many companies are based around techniques and special areas of research thus allowing synergy between researchers in the same area. For instance, in Celltech there are a number of molecular biology laboratories which carry out recombinant DNA research and hybridoma and cell culture laboratories some of which are concerned with basic work and others with scale-up and development. In summary, by creating some of the characteristics of university labs within a commercial setting the NBFs have generated a highly creative environment.

The challenge for the established companies is to try to replicate this creative environment within their research teams and divisions. A number of established companies have formed agreements with universities which involve the training of company scientists within university laboratories. However, R & D is often centralized within pharmaceutical companies and this could be a disadvantage in attempting to replicate the NBF 'culture'.

In conclusion, a high capacity for innovation is a fundamental prerequisite for success in the biotechnology industry. This can depend to a significant degree on the corporate culture as well as levels of R & D expenditure.

9.5 Global prespective

Global industries are those in which manufacturing, marketing and planning are integrated on a transnational basis so that competitive position in a given country is affected by the competitive position in others. Global industries are distinguished from multidomestic industries by the relative autonomy of the latter in each national market. Examples of global industries are automobiles, computers and pharmaceuticals. Ohmae identifies four factors contributing to the tendency towards globalization [96]:

- the growing capital intensity of manufacturing and R & D which leads to larger economies of scale;
- the accelerating rate of technological change and diffusion which means that companies have simultaneously to introduce products in the United States, Europe and Japan rather than sequentially;
- the emergence of 'global consumers' as a result of mass media and mass travel;
- the imposition of neo-protectionist measures.

Parts of the biotechnology industry such as pharmaceuticals will be global because biotechnology innovations are occurring within industries which are already global. Biotechnology companies involved in pharmaceuticals will need the capability for competing in the United States, Europe and Japan. Previously, global industries have only emerged after a long developmental period. The biotechnology industry is undergoing a simultaneous global emergence, resulting as we have seen from the world-wide diffusion of existing bioprocessing capability and molecular biology information. This should lead to an acceleration in the globalization of certain sectors of the biotechnology industry. The pharmaceutical industry, with large economies of scale in R & D and in marketing, will be the major area to experience globalization of biotechnology. Some other areas such as food, specialty chemicals and plant propagation are not likely to experience this effect to the same extent.

Distribution

American and European pharmaceutical companies already have distribution channels for world-wide distribution and competition. For instance, Warner Lambert, the American company, has 53 per cent of its sales in the United States and 45 per cent overseas, while Ciba-Geigy has 2 per cent of its sales in its home country (Switzerland) and 98 per cent in other countries. These companies are, therefore, already well positioned to introduce biotechnology products into any market or all markets simultaneously in order to achieve maximum economies of scale.

American and European NBFs are not multinational companies and do not possess the capability for global competition. Some NBFs do have overseas subsidiaries for marketing purposes, for example, Hybritech has a subsidiary in Belgium and Genentech has one in Japan. Biogen has operational facilities in both Switzerland and the United States, and both Centocor and Molecular Genetics are opening subsidiaries in Holland [48]. Most NBFs, including those with overseas subsidiaries, rely on foreign companies to distribute their products (Table 9.2).

From the Japanese perspective, the key question is how to gain access to foreign markets especially in the United States. A study by the Japanese Productivity Centre in 1982 estimated that the cost of establishing a subsidiary would be about $80m over four years and concluded that Japanese companies should form joint ventures with

Table 9.2 Some marketing agreements between NBFs and established Japanese firms

NBF (US & Europe)	Japanese	Agreement
Enzo Biochem	Meiji Seika Kaisha	World marketing of Enzo pregnancy kit
Genex	Takara Shuzo	Marketing of Genex linkers
Hybritech	Mitsubishi Chemical	Marketing of Hybritech products in Japan
Celltech	Sumitomo	Japanese marketing of Celltech diagnostics and research services
Integrated Genetics	Fujirebro	Japanese manufacture and distribution of Integrated hepatitis B diagnostic test

Source: Abstracts in Biocommerce. cumulative Vol. 1–6. 1984

American companies. By 1984 only two such joint ventures had been established, namely, Takeda with Abbott and Fujisawa with Smith/Kline.

Although Japanese companies introduced seventeen new drugs in 1983 as against six in the United States (OTA), the Japanese industry is fragmented. There are 385 companies involved in pharmaceuticals none of which represents more than about 5 per cent of the market (Table 9.3). Only one Japanese company, Takeda is among the top twenty companies in the world and this company has only 6 per cent of its sales outside Japan. Japanese companies represent 17 per cent of world pharmaceutical R & D expenditure (as against 28 per cent in the United States) and the spending on biotechnology is relatively low in comparison with American NBFs which in turn is smaller than American established firms (Table 9.4). This would suggest that Japanese companies may not be as successful in international penetration of biotechnology-health care markets as they have been in electronics, at least in the near future.

Access to technology

Because biotechnology pharmaceutical markets will be global, technological threats and opportunities may arise from any country. Firms will need to identify and act on international technological developments and to engage in international technology transfer. They will need to be able, that is, to engage in licensing from companies in various countries and to identify and form agreements with relevant research teams. Companies will require a 'presence' in each of the three major biotechnology regions, the United States, Europe and Japan. Thus American companies, will have subsidiaries or offices in Europe and Japan, Japanese in United States and Europe, and so on. Those subsidiaries may engage in manufacturing or distribution but one of their main functions will be to engage in environmental scanning and accessing of technology.

Table 9.3 Percentage share by leading companies of Japanese pharmaceutical market. 1983

Company	Share
Takeda	5.6
Shionogi	5.3
Fujisawa	4.7
Sankyo	4.1
Eisai	3.5
Taiho	2.9
Green Cross	2.6
Yamanouchi	2.5
Meiji Seika	2.5
Daiichi	2.5
Chugai	2.3
Tanabe	2.3
Kyowa Hakko	2.2
Toyo Jozo	2.0
Otsuka	1.9
Pfizer	1.7
Banyu	1.5
Toyama	1.5
Mochida	1.5
Bristol-Myers	1.4

Source: Financial Times, 17 April 1984
Note: Reproduced by permission

Table 9.4 Annual biotechnology R & D expenditure for some leading Japanese companies 1983

Company	Biotechnology R & D expenditure (US$m)
Takeda	3.6
Suntory	1.2
Sapporo Breweries	1.2
Mitsubishi Chemical	3.2
Teijin	2.4
Toyo Jozo	0.9
Kyowa Hakko Kogyo	35.2
Kirin Brewery	1.2

Source: Compiled from BIDEC. *Japan Bio Industry Letters*. April 1984

9.9 Financing

The availability of finance is not a problem for established firms diversifying into biotechnology. Such firms can use their large existing earnings and/or debt to finance the cost of biotechnology R & D. investment in manufacturing plant or distribution. NBFs on the other hand. face major financial constraints and access to adequate financing will be crucial to their future. Many NBFs have gone through a number of phases of financing. Usually a company is founded by venture capital in a number of rounds of financing. Genentech. for instance. received $.1m from Kleiner and Perkins in its first-round venture capital funding and this was followed by $10m from Lubrizol. Many of the leading NBFs then 'went public' and raised large sums through public share offering. As the share prices for biotechnology companies have fallen in the last two to three years. companies have begun to rely on R & D limited partnerships to raise additional funding especially for financing clinical trials. Recently. there has been some evidence of investor reluctance to invest in R & D limited partnerships as exemplified by Centocor which raised only $15m in a R & D limited partnership offered in 1984. Data prepared by the Office of Technology Assessment shows the rate at which NBFs are using their funds. Table 9.5 shows that the drop in working capital is large compared to

Table 9.5 Cash drain relative to equity for six new biotechnology firms in the United States, Fiscal Year 1982 ($m)

New biotechnology firm	Equity capital	Cash flow*	Yearly change in working capital	Cumulative deficit
Biogen	61.9	(3.0)	(12.1)	10.0
Cetus	128.3	5.7	(15.7)	(0.3)
Genex	13.3	0.6	(9.4)	(2.3)
Genentech	53.1	1.0	11.4	(0.03)
Hybritech	17.6	(4.3)	(6.3)	(12.8)
Molecular Genetics	1.5	(3.6)	(1.6)	(4.0)

* Cash flow is sum of net income or loss plus non-cash expenses such as depreciation
Source: OTA, based on information from company annual reports

their equity capital. Genentech, for instance, 'used up 21 per cent of its ending equity capital in 1981' (OTA).

The transition of the NBF into a fully integrated firm involves major (and probably sudden) changes in the nature of its operating revenue. At present, most NBF operating revenue is composed of contract R & D revenue, royalty payments from licenses and interest income. As the company's major products come on stream, product revenues come to comprise the largest proportion of the income. This can occur only if the firm has been successful in financing regulatory approval as well, of course, as financing the manufacturing operation. Many NBFs in the health care area, however, will find it very difficult to raise adequate funding for the clinical trials of biotechnology-based drugs. There would appear to be an incompatibility between their strategic goals of becoming drug manufacturers and the resources presently or conceivably in future available to them. Many of these smaller health care-oriented NBFs are likely to be acquired. NBFs which have targeted less expensive areas such as food, specialty chemicals and agriculture may enjoy more success in the transition to independent manufacturing operations. Interestingly, of the original 'big four' NBFs, the one which concentrates on specialty chemicals. Genex, is the most advanced in the process of forward integration.

Conclusion

The strategies described are for entry into the emerging biotechnology markets. Since a range of new biotechnology products is only beginning to be introduced, it is too early to assess the competitive behaviour of firms in the marketplace. In five years time, when biotechnology markets have developed further, it would be useful to carry out case studies on competitive strategy in various markets. Table 3.4 which lists groups of firms developing the same product, might provide a useful starting point for such an analysis.

Some of the general factors affecting success in the biotechnology industry have been described here; the nature of the markets, the technologies, and the strategic choices and constraints affecting the NBFs and established firms. At a later stage in the development of the industry, much more data should be available on the empirically observed key success factors in various biotechnology markets. For instance, what factors will be found to be related to the market success of an NBF competing against an established firm in a pharmaceutical market or to the success of an established firm diversifying into plant biotechnology.

Another fruitful area for future investigation is organizational structure and its links with competitive ability. What organizational structure will NBFs have after forward integration and acquisition of capacity for global competition? In this position NBFs may not simply copy the structure of existing multinational enterprises but may generate innovative structures. A number of leading multinational enterprises are now undergoing a major structural reorganization to promote innovation and entrepreneurship within their divisions. Thus established biotechnology firms and NBFs may both pioneer structural innovations as well as new technologies. Related to this is the question of the role of the R & D activity within NBFs after forward integration. Currently R & D represents 70–95 per cent of NBF 'sales'. After forward integration, there are two possibilities; the R & D expenditure as a percentage of sales may decline to match the industry average (10–12 per cent for pharmaceuticals) or alternatively it could remain high (20 per cent) thus making biotechnology a much more research intensive industry even than general pharmaceuticals.

An extensive literature has arisen in recent years on the electronics industry, largely due to its high growth rate and innovativeness which act as a model for innovation in high technology. Over the next ten years the biotechnology industry should provide a similar source of analyses which may have normative aspects beyond biotechnology.

Bibliography

1. Arber. W.. 1965. *Ann. Rev. of Microbiol*. vol. 19. 365–78.
2. Backman. K. *et al*. 1984. Use of synchronous site specific recombination *in vivo* to regulate gene expression. *Bio/Technology*. 1045–9. December 1984.
3. Barstow. Leon E.. 1984. Automated chemical synthesis of peptides and DNA fragments. *Bio/Technology*. vol. 2. no. 5. 445. May.
4. Bio-Response Inc. Prospectus. 12 January 1983.
5. *Bio/Technology*. 1983. W.R. Grace – a slow moving giant leaps into bioresearch. Vol. 1. no. 1. 30–1. March.
6. *Biotechnology News*. 1983. Genex shifts research. seeks additional financing. 1 April.
7. *Biotechnology News*. 1983. Cetus Madison has successfully engineered a plant to have a desirable property. 15 February.
8. *Biotechnology News*. 1983. Farley leaves Cetus to form new company. 15 May.
9. *Biotechnology News*. 1983. Cetus' first diagnostic test is now on the market. 15 March.
10. *Biotechnology News*. 1983. Centocor. 22 February.
11. *Biotechnology News*. 1984. Chiron produces interleukin-2 via yeast. 1 January.
12. *Biotechnology News*. 1984. DNA hybrid probes: the business begins to take shape. 15 March.
13. *Biotechnology News*. 1984. Wellcome licenses Biogen's hepatitis B vaccine. 1 November.
14 *Biotechnology News*. 1984. More clinical trials start as IL-2 race becomes more intense. 15 July.
15. *Biotechnology News*. 1984. Collagen and Monsanto sign deal on growth factors. 1 May.
16. *Biotechnology News*. 1984. Invitron to commercialize Monsanto's cell culture technology. 1 October.
17 *Biotechnology News*. 1984. Monsanto and Oxford to study sugar molecules. 1 January.
18. *Biotechnology Newswatch*. 1982. Genentech. Speywood close ranks in move to clone Factor VIIIc. 15 November.
19. *Biotechnology Newswatch*. 1984. Monsanto. 5 November.

20. *Biotechnology Law Report.* 1984. Second Cohen-Boyer patent covering spliced genes issues. Vol. 3. no. 9-90. 183-95.
21. Bryan. Jenny. 1984. British drug giants move into biotech. *Bio/ Technology.* vol. 2. no. 5. 388-95. May.
22. Bull. A.T.. Holt. G. Lilly. M. 1982. Biotechnology - international trends and perspectives. *OECD.*
23. Bull. Daniel N.. 1983. Fermentation and genetic engineering: problems and prospects. *Bio/Technology.* vol. 1. no. 10. 847-56. December.
24. *Business Week.* 1984. The biotech bigshots snapping up small seed companies. 11 June.
25. *Business Week.* 1984. The next revolution in medicine is almost here. 8 October.
26. California Biotechnology Inc.. 1983. Prospectus. 26 October.
27. Cetus Corporation. 1981-1984. Annual Reports.
28. Celltech Annual Reports. 1981-1984.
29. Centocor Inc. Annual Reports 1982-1984.
30. Channon. Derek E.. Jalland. Michael. 1979. *Multi-national strategic planning.* Macmillan. London.
31. Collaborative Research Inc.. 1982. Prospectus. 11 February.
32. Edwards. Christopher. 1983. Industrial Policy: Where is biotech? *Bio/ Technology.* vol. 1. no. 10. 821. December.
33. Edwards. Christopher. 1983. American biotechnology needs a strategic planning center. *Bio/Technology.* vol. 1. no. 1. 7 March.
34. Edwards. Christopher. Elkington. G.J.. Murray. A.M.. 1984. Japan taps into new biotech. *Bio/Technology.* vol. 2. no. 4. 307-21. April.
35. Eli-Lilly. 1981. Press release. First genetically engineered human insulin trials. 23 June.
36. Eli-Lilly. 1982. Press release. Eli-Lilly announces approval of human insulin (recombinant DNA origin) by United Kingdom licensing authority. 17 September.
37. Eli-Lilly. 1982. 1983. Annual reports.
38. Ellis. P.B.S.. 1982. Biotechnology. Industry emergence. development and change. M.Sc. dissertation. MII. June.
39. *European Chemical News.* 1981. Celltech plans rapid expansion joint ventures and spin off companies. 28 September.
40. *European Chemical News.* 1982. Celltech to market two more mono-clonals this year. 25 January.
41. *European Chemical News.* 1983. Monsanto in plant genetics first. 7 February.
42. *European Chemical News.* 1984. Biotechnology Supplement. May.
43. *European Chemical News.* 1984. Novo plans human insulin production plant. 12 November.
44. *European Chemical News.* 1984. Biotechnica to patent process. 10 September.

45. *European Chemical News*. 1984. Joint venture in seeds research. 26 November.
46. *European Chemical News*. 1984. Genentech seeks cash for anticancer drug. 3 December.
47. *European Chemical News*. 1984. Cetus. Grace finalize joint venture. 10 September.
48. *European Chemical News*. 1984. Centocor plans Mab plant in the Netherlands. 17 December.
49. *Financial Times*. 1981. Biogen finds it easy to raise money in Europe. 23 October.
50. *Financial Times*. 1981. Genentech – snug in the incubator. 3 December.
51. *Financial Times*. 1982. Celltech: gene machine coming on stream. 16 June.
52. *Financial Times*. 1982. Big plans for small biotech. 31 March.
53. *Financial Times*. 1982. Lilly replies to Novo's drug shot. 15 June.
54. *Financial Times*. 1983. Human insulin fights for a market share. 28 April.
55. *Financial Times*. 1984. Why Genentech wants to grow up. 18 April.
56. Freeman. Christopher. 1974. *The economics of industrial innovation*. Penguin. London.
57. Freeman. Karen. 1984. Eli-Lilly's $60m biomedical research centre to open in September. *Genetic Engineering News*. July/August.
58. Genentech. 1981–84. Annual reports.
59. Genentech. 1976–84. Summary of press releases.
60. *Genetic Engineering News*. 1984. Centocor calmly weathers criticisms and competition in cancer diagnostics. vol. 4. no. 7. October.
61. *Genetic Engineering News*. 1984. Third annual GEN guide to biotechnology companies. vol. 4. no. 8. November/December.
62. *Genetic Engineering News*. 1984. Cohen-Boyer plasmid patent issues: many questions remain. Vol. 4. no. 7. October.
63. *Genetic Engineering News*. 1984. Symbolic nature of biological data makes artificial intelligence attractive to biotech. Vol. 4. no. 6. September.
64. Genex Corp. 1982–84. Annual Reports.
65. Genex corp. 1984. 83 Press releases. 1984. 6 August and 7 May: 1983. 16 December. 4 November and 15 August.
66. Goeddel. D.V. *et al.*. 1979(a). Direct expression in *Escherichia coli* of a DNA sequence coding for human growth hormone. *Nature* 281. 544–8.
67. Goeddel. D.V. *et al.*. 1979(b). Expression in *Escherichia coli* of chemically synthesized genes for human insulin. *Proc. nat. Acad. Sci.*. 76. 106–10.
68. Goeddel. D.V. *et al.*. 1980. Human leukocyte interferon produced by

E. coli is biologically active. *Nature.* 287. 411-16.

69. Gordon. A. *et al.*. 1984. Characterization studies on human melanoma cell tissue plasminogen activator. *Bio/Technology.* 1051-7. December.

70. Glick. J. Leslie. 1981. The cost reducing effects of new medical technologies: health care in the age of recombinant DNA and applied genetics. Submitted to Institute of Health Economics and Technlogy assessment. 27 November.

71. Harris. T.J.R. *et al.*. 1982. Molecular cloning and nucleotide sequence of DNA coding for prochymosin. *Nucleic Acids Research.* 10. 2177-87.

72. Hayes. R.. Abernathy. W. 1980. Managing our way to economic decline *Harvard Business Review.* 67-77. July/August.

73. Horowitz. Mel. Sakakibara. K. 1983. The changing strategy-technology relationship in technology-based industry: a comparison of the United States and Japan. Paper to Annual conference of the Strategic Management Society. Paris. 27-29 October.

74. Integrated Genetics. 1983. Prospectus. 26 May.

75. Johnson. Chalmers. 1984. The industrial policy debate re-examined *California Management Review.* Vol. 28. no. 1. Fall.

76. Jones. H.. Twiss. Brian C.. 1978. *Forecasting technology for planning decisions.* Macmillan. London.

77. Kelly. T.J.. Smith. H.O.. 1970. *J. Mol. Biol.*. 51. 393-409.

78. Klausner. Arthur. 1983. Upjohn consolidates biotech efforts into a single unit: plans expansion. *Bio/Technology.* vol. 1. no. 10. 837. December.

79. Klausner. Arthur. 1984. Firms seek to score on protein analysis. *Bio/ Technology.* vol. 2. no. 12. 1011-12. December.

80. Klausner. Arthur. 1984. Flow of biotech products just a trickle? *Bio/ Technology.* vol. 2. no. 6. 499-501. June.

81. Klotz. Lynn. C.. 1984. Is genetic engineering just another South Sea bubble? *Bio/Technology.* January.

82. Knickkrehm. Glenn. 1984. How to value your biotechnology company. *Genetic Engineering News.* vol 4. no. 8. 42-3.

83. Lasky. L.A. *et al.*. 1984. Protection of mice from lethal herpes simplex virus infection by vaccination with a secreted form of cloned glycoprotein D. *Bio/Technology.* 527-38. June.

84. Lederman. Leonard. 1984. 'Foresight activities in the USA: time for a reassessment'. *Long Range Planning.* vol. 17. no. 3. 41-51. June 1984.

85. Luckin. P.A. *et al.*. 1984. Molecular cloning of AIDS associated retrovirus. *Nature.* vol. 312. 768-71.

86. McCormick. Douglas. 1984. Microcomputer controls for biotechnology *Bio/Technology.* vol. 2. no. 12. 1022-8. December.

87. Maidique. Modesto A.. 1983. Point of view: The new management thinkers. *California Management Review.* vol. *xxvi.* no. 1. Fall.

88. Mason. F.A. *et al.*. 1984. Purification of calf prochymosin (prorennin)

synthesized in *Escherichia coli. Bio/Technology* September.
89. Monoclonal Antibodies Inc.. 1982. Prospectus. 21 December.
90. Monsanto Co.. 1982. 1983. Annual reports.
91. Mowery. David. C.. 1983. Innovation. market structure and government policy in the American semiconductor electronics industry: A survey. *Research Policy*. 12. 183–97.
92. Murray. James. R.. 1983. Financing for health applications increases
93. *Nature*. 1982. Monsanto hands out $23.5 million. Vol. 297. 17 June.
94. *OECD*. 1979. The impact of multinational enterprises on national scientific and technical capacities: pharmaceuticals DSTI/SPR/79.10-MNE. Addendum to document DSTI/SPR/77.34-MNE.
95. Ohmae. Kenichi. 1982. *The mind of the strategist*. Penguin. Harmondsworth. Middx.
96. Ohmae. Kenichi. 1985. *Triad power*. Free Press. New York.
97. Osawa. Toshiaki. 1984. Human T-cell hybridomas producing inflammatory lymphokines. *Trends in Biotechnology*. vol. 2. no. 2.
98. *Office of Technology Assessment* (US Congress). 1981. Impacts of applied genetics.
99. *Office of Technology Assessment* (US Congress). 1984. Commercial biotechnology: An international analysis.
100. Old. R.W.. Primrose. S.B.. 1980. *Principles of gene manipulation*. Blackwell Scientific Publications. Oxford.
101. Pascale. P.. Athos. A.. 1981. The art of Japanese management: Applications for American executives. Warner. New York.
102. Peters. Thomas J.. Waterman. Robert A. Jr.. 1982. *In search of excellence. Lessons from America's best-run companies.* Harper and Row. New York.
103. Porges. Amelia.. 1984. Biotechnology trade with Japan. *Bio/Technology*. vol. 2. no. 4. 327-32.
104. Porter. Michael. 1980. *Competitive strategy*. Free Press. New York.
105. Porter. Michael. 1983. *The technological dimension of competitive strategy*. In *Research on technoligical innovation, management and policy*. Vol. 1. 1-33. JAI Press.
106. *Practical Biotechnology*. 1983. Novo. September.
107. Price. J.D. de Solla.. 1965. Is technology historically independent of science? *Technology and Culture*. vol. vi. no. 4.
108. Schmookler. J.. 1966. *Invention and economic growth*. Harvard University Press. Cambridge (Mass.).
109. Sherman. Philip M.. 1982. *Strategic planning for high technology industries.* Addison-Wesley. 1982.
110. Slatter. Stuart St. P.. 1970. *Competition and marketing strategies in the pharmaceutical industry.* Croom Helm. London.
111. Treble. M. J.. 1982. *Genetic Engineering News*. July/August. Quoted by OTA Report [99] p. 150.

112. Ulmer. K.. 1983. Protein engineering. *Science*. vol. 219. 11 February.
113. United States Department of Commerce. 1984. *High technology industries profiles & outlooks: Biotechnology*.
114. Vane. J.. Cautrecasas. P.. 1984. Genetic engineering and pharmaceuticals. *Nature*. vol. 312. 303–5.
115. *Vournakis. J.N.. Elander. R.P.. 1983. Genetic manipulation of antibiotics producing microorganisms. Science*. 219. 703–9.
116. Weitz. Barton A.. Wensley. Robin. 1984. *Strategic marketing-planning implementation and control*. Kent Publishing Co.. Boston.
117. Wilson. Tazewell. Klausner. A.. 1984. Computers reveal proteins mysteries. *Bio/Technology*. vol. 2. no. 6. 511–19.
118. Wilson. Tazewell. 1984. More proteins from mammalian cells. *Bio/Technology*. vol. 2. no. 9. 753–5.

Glossary of technical terms

Amino Acid Organic acids which constitute the building blocks of proteins. They are produced industrially for use in food. feed and pharmaceutical industries.

Antibody A protein produced by living organisms in response to a foreign agent or antigen.

Antigen A protein or other molecule which when injected into a human or animal body will generate the productin of an antibody.

'Aspartame' An alternative sweetener marketed by G.D. Searle. It is composed of two amino acids. phenylalanine and aspartic acid.

Bacteriophage A type of virus which replicates inside bacteria.

B cell Cf. lymphocyte.

B cell growth factors (BCGFs) Substances which cause proliferation and differentiation of B lymphocytes.

Biochemical engineering The design and operation of unit processes and systems in industrial fermentation.

Bioinformatics The area of convergence between biotechnology and information technology. It includes such areas as software for use in genetic engineering research.

Biopolymer Any natural substance composed of a series of smaller molecules. Used here it refers to microbial polysaccarides.

Calcitonin A small peptide hormone produced by the thyroid which regulates calcium transport and uptake.

Cell culture The growth of cells obtained from higher animals or plants *in vitro*.

Cloning The replication of a single cell line. Molecular cloning refers to recombinant DNA technology.

Collagen A protein found in bone and connective tissue.

Commodity chemical Chemicals produced in large volume but with low value. Examples include industrial solvents such as acetone. alcohol, butanol, organic acids such as citric and chemical feedstocks for plastics. synthetic rubber or pharmaceutical industries.

144

Cytomegalovirus   A member of the herpes virus group which can cause disease in children and those having blood transfusions.

Diagnostics (*in vitro*)   Diagnostic kits and systems for use on tissue or fluid samples in the laboratory. Included here are tests which have been available for some time and also new tests incorporating monoclonal antibodies.

Diagnostics (*in vivo*)   Diagnostic technology for use within the body such as monoclonal antibody-based visualization of cancer cells.

DNA   Deoxyribonucleic acid. The chemical out of which genes are made.

Enzyme   A protein which acts as a catalyst in biological reactions.

Enzymeimmunoassay   A diagnostic method which employs ensyme-labelled antibodies.

Erythropoietin   A protein involved in regulating production of red blood cells. It may have therapeutic uses in treating anaemia in patients with chronic kidney disease.

Escherichia coli (E. coli)   A species of bacterium which lives in the intestinal tract of man and other vertebrates. It is widely used as a host for recombinant DNA work.

Expression system   A construction of DNA incorporating the gene coding for a particular substance which has been made so as to maximize production of that substance within the cell.

Factor VIII   A protein involved in blood clotting which is used in the therapy for haemophilia.

Fermentation   A process which uses living micro-organisms (or their products) to cause a desired chemical transformation of a particular substance.

Gene   The fundamental unit of heredity. It is composed of the chemical DNA and the information contained in its structure is used in the cell to direct the production of a particular protein.

Genetic engineering   (also recombinant DNA or rDNA). The construction and manipulation of hybrid DNA to introduce genes coding for desired proteins into specific organisms.

Genome   The total genetic information within an individual.

Glycosylation   The chemical addition of a sugar molecule to another molecule, usually a protein.

Growth hormone   A peptide involved in the regulation of growth.

Human chorionic gonadotropin   A glycoprotein hormone

produced by the placenta and involved in controlling progesterone secretion by the corpus luteum. It forms the basis of some diagnostic tests for pregnancy.

Human serum albumin The major protein component of human plasma. It is used medically for treating shock, burns and in some types of surgery.

Hybridoma A hybrid cell used for the production of monoclonal antibodies. It is formed from the fusion of a myeloma cancer cell and an antibody producing B lymphocyte.

Immunoassay Any diagnostic test incorporating the use of antibodies either polyclonal or monoclonal.

Immunomodulator A class of substances which regulate the activity of the immune response.

Immunosuppressant A class of substances which cause suppression of the immune response.

Immunotoxin A cell poison (such as ricin) which is targeted against tumour cells by use of a specific antibody.

Interferons A class of immune regulators or lymphokines which are involved in the responses of cells to viral infection and cancer. There are three subgroups, alpha, beta and gamma interferons. They have received much attention as future anti-cancer agents.

Interleukin-2 A type of immunomodulator which is being tested for anti-cancer effects. It stimulates T cell growth *in vivo*.

Kidney plasminogen activator A factor which causes activation of plasmin which breaks down blood clots.

Lymphocyte White blood cells involved in the immune response. There are two main types, B lymphocytes (B cells) and T lymphocytes (T cells). The former produce antibodies; the latter are involved in cell mediated immunity and in helping B cells.

Lymphokines A class of immunoregulators produced by lymphocytes.

Lymphotoxin A lymphokine which has cytotoxic effects on tumour cells but does not effect normal cells.

Macrophage activating factor A lymphokine released from T cells and causing activation of macrophages. It is being investigated for possible anti-cancer effects.

Monoclonal antibody (Mab) A highly specific type of antibody produced by a single clone of cells which can recognize only one antigenic site.

Myeloma   A type of tumour cell. Cf. hybridoma.

Oligonucleotide   A short segment of DNA. It is often used to refer to a DNA segment artifically constructed for use in a probe or linker.

Oncogene   A gene which when activated is involved in the transformation of normal cells. Two or more type of oncogene may need to co-operate to turn a cell cancerous. Some oncogenes code for cellular growth factors or their receptors.

Peptide   A short segment of protein.

Pheromone   An externally secreted animal substance involved in effecting social or sexual signals between individuals of the same species.

Plasmid   A small loop of extrachromosomal DNA used as a vector in recombinant DNA research.

Promotor   A region of DNA which occurs close to the origin of a gene and is involved in switching it on.

Protein engineering   The study of the relationship of protein structure and function with a view to designing proteins with specific functional characteristics.

Radioimmunoassay (RIA)   A diagnostic method which employs radiolabelled antibodies.

Restriction enzymes   Enzymes which cut DNA. There are two classes; Class I enzymes cut DNA randomly while Class II enzymes cut DNA at particular target sequences.

Ricin   A cell poison which may be useful in immunotoxins for destruction of tumour cells.

Somatomedins   A group of small peptide growth factors some of which are produced in the liver in response to growth hormone.

Specialty chemical   Low-volume, high-value chemicals such as enzymes.

T cell   Cf. lymphocyte.

Tissue plasminogen activator   A substance which causes activation of plasmin which is involved in the breakdown of blood clots.

Tissue typing   The grouping of tissues into immunological classes.

Tumour necrosis factor   A macrophage produced protein which exhibits *in vitro* and *in vivo* killing of tumour cells. It has about 30 per cent homology of amino acid sequence with lymphotoxin.

Urokinase   A thrombolytic enzyme involved in breakdown of blood clots. It occurs in human urine.

Vaccine   An agent which confers active immunity. It consists of
attenuated or killed bacteria or virus or their proteins.

Vector   In recombinant DNA research it refers to any piece of
DNA such as a plasmid, phage or virus which can be used to
introduce new genes into a cell.

Virus   Intracellular parasites consisting of a core of DNA (or RNA)
surrounded by a protein coat. They infect bacterial, animal and
plant cells and cannot reproduce outside these hosts.

Index